Contents at a Glance

Table of Contents

HACKING
RASPBERRY PI®

TIMOTHY L. WARNER

800 East 96th Street,
Indianapolis, Indiana 46240 USA

Hacking Raspberry Pi®

Copyright © 2014 by Que Publishing

ISBN-13: 978-0-7897-5156-0
ISBN-10: 0-7897-5156-9

Library of Congress Control Number: 2013944701

Printed in the United States of America

First Printing: November 2013

Trademarks

All terms mentioned in this book that are known to be trademarks or service
marks have been appropriately capitalized. Que Publishing cannot attest to
the accuracy of this information. Use of a term in this book should not be
regarded as affecting the validity of any trademark or service mark.

Warning and Disclaimer

Every effort has been made to make this book as complete and as accurate
as possible, but no warranty or fitness is implied. The information provided
is on an "as is" basis. The author and the publisher shall have neither liability
nor responsibility to any person or entity with respect to any loss or damages
arising from the information contained in this book.

Bulk Sales

Que Publishing offers excellent discounts on this book when ordered in
quantity for bulk purchases or special sales. For more information, please
contact

> **U.S. Corporate and Government Sales**
> 1-800-382-3419
> corpsales@pearsontechgroup.com

For sales outside of the U.S., please contact

> **International Sales**
> international@pearsoned.com

Editor-in-Chief
Greg Wiegand

Executive Editor
Rick Kughen

Development Editor
Todd Brakke

Managing Editor
Kristy Hart

Project Editor
Elaine Wiley

Copy Editor
Chrissy White

Indexer
Brad Herriman

Proofreader
Kathy Ruiz

Technical Editor
Brian McLaughlin

Editorial Assistant
Kristen Watterson

Cover Designer
Chuti Prasertsith

Book Designer
Mark Shirar

Senior Compositor
Gloria Schurick

Graphics
Tammy Graham

Cover Illustration
©Kokander/Shutter
Stock

©Yippe/Shutter Stock

About the Author

Timothy L. Warner has helped thousands of beginners succeed with technology in business and in schools. Until recently a digital media specialist with Pearson Technology Group, he worked in various facets of information technology, including systems administration, software architecture, and technical training. He is the author of *Unauthorized Guide to iPhone, iPad and iPhone Repair.*

Dedication

To my father, Lawrence K. Warner, whose fascination with electronic gadgets fueled and inspired my own.

Acknowledgments

Thank you to all my friends at Pearson Technology Group, especially my editors Rick Kughen and Todd Brakke, my publishers Greg Wiegand and Paul Boger, my project managers Elaine Wiley and Kristen Watterson, and my copy editors/proof readers Chrissy White and Kathy Ruiz. You are a group of wonderful people, that's for sure.

With regard to the book's subject matter, thank you to Eben Upton and Gert van Loo of the Raspberry Pi Foundation (Gert, I know you say you aren't a member of the Foundation, but you're foundational to the Pi project at any rate) for your gracious assistance—you guys are brilliant! Thanks to Brian McLaughlin for his thorough technical edit of the manuscript.

Thank you to my family (Susan, Zoey, Fred, Moon, Maya, Stub, Mom, Dad, Trish, Mom H, Rick, Chelle, Don, Alex, Kevin, Kristina, and so on) for your love and support—I treasure you all.

Finally, thank you to you, my reader. Without you I would be teaching and writing into a blank void. Don't get me wrong—I love the sound of my own voice and I'm my own favorite subject. However, discussing this technology and sharing this knowledge just wouldn't be the same without you.

We Want to Hear from You!

As the reader of this book, you are our most important critic and commentator. We value your opinion and want to know what we're doing right, what we could do better, what areas you'd like to see us publish in, and any other words of wisdom you're willing to pass our way.

We welcome your comments. You can email or write to let us know what you did or didn't like about this book—as well as what we can do to make our books better.

Please note that we cannot help you with technical problems related to the topic of this book.

When you write, please be sure to include this book's title and author as well as your name and email address. We will carefully review your comments and share them with the author and editors who worked on the book.

Email: feedback@quepublishing.com

Mail: Que Publishing
         ATTN: Reader Feedback
         800 East 96th Street
         Indianapolis, IN 46240 USA

Reader Services

Visit our website and register this book at quepublishing.com/register for convenient access to any updates, downloads, or errata that might be available for this book.

INTRODUCTION

Hi! I'm Tim Warner, and I am happy to be your guide to the Raspberry Pi. My goal for this Introduction is to get you excited about this $35 credit card-sized computer that we call the Raspberry Pi.

"What in the world can you do with a $35 computer?" you might ask. Well let's have a look at a representative smattering of some popular Raspberry Pi projects, and you can answer that question for yourself:

- Picade Arcade Cabinet (http://is.gd/i4YwQ2). This is a tiny, fully-functional arcade machine.
- Pi in the Sky (http://is.gd/4niEMV). The inventor put a camera-equipped Raspberry Pi into the upper atmosphere via a weather balloon.
- Streaming Music Jukebox (http://is.gd/oqec3H). This is an inexpensive wireless music broadcasting machine.
- Raspberry Pi Keyboard Computer (http://is.gd/tvmgC8). This is a full computer packed into the form factor of a computer keyboard.
- DeviantArt Picture Frame (http://is.gd/i7ED9w). This is an interactive frame that dynamically displays artwork from the DeviantArt website.
- Pi-Powered Bitcoin Miner (http://is.gd/DrpJ7A). This tiny rig enables you to participate in Bitcoin mining, which can yield you some hefty monetary returns.
- FM Radio Transmitter (http://is.gd/tS52Yb). This is a low-power (albeit mono) portable FM radio.
- KindleBerry Pi (http://is.gd/73iVz4). This is a Pi that uses an Amazon Kindle as a monitor and a "dumb" terminal.
- Pi-Powered Motion Detector (http://is.gd/g4Okb6). This is a Pi that can detect motion and optionally take action upon that detection.
- 3D Printer (http://is.gd/Bg83jD). This is a Raspberry Pi-powered printer that can actually fabricate three-dimensional objects.
- Cheese-Powered Radio-Controlled Car (http://is.gd/ZExbWO). This radio-controlled car is controlled with a Nintendo Wii Remote and powered by ordinary slices of cheese.
- Raspberry Pi Robot (http://is.gd/367FZ5). This is a multi-tasking robot that uses the Raspberry Pi as its "brain."

- Automated Chicken Coop Door (http://is.gd/boZR6F). This is a Raspberry Pi, a relay, and a motor that opens and closes a farm chicken coop door on a schedule.
- Pi-Powered Weather Station (http://is.gd/LDbZIr). This outdoor weather sensing and reporting unit runs on USB power.
- Audio Book Player (http://is.gd/TnkcoW). The inventor made this simple, push button-operated audio book player for his grandmother's 90th birthday.
- Android Device (http://is.gd/9eLPkV). The true mark of the software/hardware hacker is to get Linux or Android running on any device. This is a Raspberry Pi that runs Android 4.0 Ice Cream Sandwich.

I think you'll agree that the diminutive Raspberry Pi has quite a bit of potential for a tiny little circuit board. Let's take a look at how I've organized this book to provide you with maximum learning in minimum time.

The title of this book contains the inflammatory term *hacking*. This word has several meanings even within the information technology industry. First, hacking means to use or adapt a piece of electronics for a purpose other than which it was originally intended. For instance, you can hack your eReader such that it runs Linux and acts as a web server. Would you want to? Believe it or not, people do exactly that, often just to see if they can.

Second, hacking means to break into someone else's computer system, often illegally and without the owner's permission. Of course, *Hacking Raspberry Pi* embraces the first definition of the term!

What's in This Book

Hacking Raspberry Pi is organized in such a way that I gradually immerse you into software and hardware engineering. Believe it or not, configuring hardware and mastering the Linux command line isn't as difficult as you might have imagined. This book takes you through it, beginning with...

- **Chapter 1—What Is the Raspberry Pi?**: Here I explain exactly what the Raspberry Pi is, how it is important, and why you would want to learn more about the device.
- **Chapter 2—Hardware Components Quick Start**: In this chapter you become familiar with the Raspberry Pi's form factor; in other words, its onboard hardware.
- **Chapter 3—A Tour of Raspberry Pi Peripheral Devices**: What do you plug into the Raspberry Pi and where? This chapter gives you everything you need to know to get your Raspberry Pi up and running.
- **Chapter 4—Installing and Configuring an Operating System**: In this chapter you learn how to install Raspbian, the reference Linux distribution for Raspberry Pi.

- **Chapter 5—Debian Linux Fundamentals—Terminal**: This chapter is an excellent jump-start for those who always wanted to understand something of the Linux command-line environment.

- **Chapter 6—Debian Linux Fundamentals—Graphical User Interface**: Sometimes it is plain easier to navigate in Linux from a GUI environment. Here you learn how to use LXDE, the reference GUI shell in Raspbian.

- **Chapter 7—Networking Raspberry Pi**: In most cases, you'll want to put your Raspberry Pi on your local area network (LAN), if not the Internet. You learn how to configure both wired and wireless Ethernet by reading the material in this chapter.

- **Chapter 8—Programming Raspberry Pi with Scratch—Beginnings**: Scratch provides an easy-to-learn platform for learning how to develop computer programs. In this chapter you become familiar with what Scratch is and how the platform works.

- **Chapter 9—Programming Raspberry Pi with Scratch—Next Steps**: Here you develop your first real Scratch application, all on the Raspberry Pi!

- **Chapter 10—Programming Raspberry Pi with Python—Beginnings**: The "Pi" in Raspberry Pi is actually a reference to the Python programming language. Therefore, it is imperative that you learn a thing or two about programming in Python.

- **Chapter 11—Programming Raspberry Pi with Python—Next Steps**: Many Raspberry Pi projects involve one or more Python scripts. Therefore, the more experience you obtain with the language, the better.

- **Chapter 12—Raspberry Pi Media Center**: This section of the book is focused on applying your new Raspberry Pi skills to several practical projects. Here you build a Pi-powered media center running Xbox Media Center (XBMC).

- **Chapter 13—Raspberry Pi Retro Game Station**: Who does not love retro video games? In this chapter you learn how to convert your Raspberry Pi into a mobile classic game station. This is my favorite project in the entire book.

- **Chapter 14—Raspberry Pi *Minecraft* Server**: In this chapter you learn how to install, configure, and play *Minecraft* Pi Edition. You also learn how to set up your Pi as a Minecraft server.

- **Chapter 15—Raspberry Pi Web Server**: Here you configure your Raspberry Pi to serve up web pages on your own local network and/or the public Internet.

- **Chapter 16—Raspberry Pi Portable Webcam**: Many people are interested in making their own security camera or general-purpose webcam. It is surprisingly easy to do this with a Raspberry Pi; you can use the Raspberry Pi Camera Board or your own webcam.

- **Chapter 17—Raspberry Pi Security and Privacy Device**: You can use the Raspberry Pi to increase your online security and privacy, say, when you access the Internet through a public Wi-Fi hotspot, hotel room, and so forth. You learn a lot of cool stuff in this chapter!

- **Chapter 18—Raspberry Pi Overclocking**: Although the Raspberry Pi is small and inexpensive, the Raspberry Pi Foundation gives users a great deal of flexibility in squeezing every bit of performance from the device. In this chapter you learn what your options are and how to leverage them to customize the behavior of your Pi.

- **Chapter 19—Raspberry Pi and Arduino**: For my money, the combination of the Raspberry Pi and the Arduino is unbeatable with Pi's flexibility and the Arduino's singleness of purpose. In this chapter you use the Arduino Uno and Alamode with your Pi and get some great project ideas.

- **Chapter 20—Raspberry Pi and the Gertboard**: I close the book by teaching you how to use the ultimate Raspberry Pi add-on board, the Gertboard. The Gertboard is a kitchen sink expansion board that provides you with many opportunities for experimentation and learning.

This book is chock-full of tasks that give you guided experience at setting up, configuring, troubleshooting, and building projects with your Raspberry Pi. I strongly suggest you work through as many tasks as possible.

I am confident that by the time you finish this volume, you will not only be able to discuss the Raspberry Pi intelligently, but you'll also have an excellent baseline familiarity with practical, applied computer science.

Who Can Use This Book

Ah, now we come to the "Exactly who is this book intended for?" question. Actually, I have a very detailed view of those of you who will benefit most from this book:

- **Students and Teachers**: The Raspberry Pi was developed by educators for educators and their students. Due to its open architecture and low price point, people can use the Pi as a platform for learning how computer hardware works at a low level with minimal risk. After all, if the worst happens and you fry your Pi, your investment loss is limited to $25 or $35.

- **Hardware and Software Hackers**: As I discussed earlier in this Introduction, hacking has myriad goals. Here I refer to those who want to leverage the Pi to accomplish some business or personal goals, with or without the addition of third-party extension hardware and software.

- **Tech Enthusiasts**: These people are do-it-yourselfers (DIYers) who are of the mind, "It's cheaper for me to make it myself," or better yet, "I can make this better than anything I can buy." If you are among these individuals, then kudos to you! You are among a small elite.

How to Use This Book

I hope this book is easy enough to read that you don't need instructions. That said, a few elements bear explaining.

First, I love to provide relevant websites, but as you know, some URLs are absurdly long and difficult to transcribe. To that end, I make use of the is.gd (http://is.gd) URL shortening service. I hope you find my is.gd "shortie" URLs convenient. One important note about those is.gd URLs: they are case-sensitive, so if you type the URL http://is.gd/6zwzwT as http://is.gd/6ZWZWT or some other variation, the link will not work correctly.

Second, this book contains several special elements, presented in what we in the publishing business call "margin notes." There are different types of margin notes for different types of information, as you see here.

NOTE

This is a note that presents information of interest, even if it isn't wholly relevant to the discussion in the main text.

TASK: THIS IS A TASK

This is a step-by-step procedure that gives you practice in whatever technology we're discussing at the moment. Almost every chapter in this book has at least a couple tasks for you to perform that will help you get the most out of your Raspberry Pi.

There's More Online...

When you need a break from reading, feel free to go online and check out my personal website, located at timwarnertech.com. Here you'll find more information about this book as well as other work I do. And if you have any questions or comments, feel free to send me an email at tim@timwarnertech.com. I do my utmost to answer every message I receive from my readers. Thanks very much for reading my book, and I hope that it exceeds your expectations!

## CHAPTER 1

What Is the Raspberry Pi?

The Raspberry Pi, with Pi pronounced *pie*, is a $35 personal computer about the size of a credit card. No kidding—the Raspberry Pi development team literally used a credit card as a template when they designed the Pi's printed circuit board (PCB).

The Pi, or RasPi, or RPi (users enjoy creating nicknames for the device) is the brainchild of the Raspberry Pi Foundation (http://raspberrypi.org), a charity based in the United Kingdom (UK) and founded by Broadcom hardware architect Eben Upton, along with some of his esteemed associates at Broadcom, Cambridge University, and other corporate and educational organizations.

Historically, computer science curricula both at the K-12 and even collegiate levels—if there is any curricula at all—tend to focus more on theory than on practical application of computing concepts. Eben and the Foundation conceived the Raspberry Pi in 2006 as a way to make computer science more accessible to students.

Eben and I arose from the same generation of programmers, which is to say we came of age during the early 1980s and cut our teeth learning Beginner's All-Purpose Symbolic Instruction Code (BASIC) programming on microcomputer platforms such as the Amstrad CPC, Commodore 64, Tandy TRS-80, and others.

The Raspberry Pi was intended to be the cultural successor to the Acorn BBC Micro personal computer that was extremely popular in the UK during the 1980s (see Figure 1.1).

FIGURE 1.1   The BBC Micro personal computer of 1981 was the prototype for the Raspberry Pi of 2012.

The BBC Micro shipped with a 2 MHz MOS Technology 6502 central processing unit (CPU). Later models, specifically the Archimedes, introduced the Acorn RISC Machines (ARM) processor. Believe it or not, the ARM processor platform is still alive and well in the 21st century; its application is mainly targeted to mobile phones and tablet computers. The Raspberry Pi is equipped with an ARM1176JZF-S (often abbreviated as ARM11) CPU; we'll learn much more about Pi hardware in Chapter 2, "Hardware Components Quick Start."

NOTE: AN ARM AND A LEG

For CPU devotees in my readership, allow me to tell you that the ARM11 is a 32-bit microprocessor that uses the Reduced Instruction Set Computing (RISC) processing architecture.

Why the Pi?

Many consider the Raspberry Pi to be an ideal platform for teaching both kids and adults how computer science works because it requires minimal investment. Any interested individual can learn not only how to program computer software, but also to work directly with electronics and computer hardware. If an experiment goes wrong and the Pi becomes inoperable, then the student is out only $35 as opposed to hundreds or thousands.

My use of the term *computer science* is intentional. What's so cool about the Raspberry Pi is that we can move beyond surface-level software and interact directly with the internals of what most people consider to be a "black box." A Raspberry Pi-based education can form the foundation of sought-after hardware and software engineering skills, which are lucrative and extraordinarily valuable in today's global job marketplace.

To be sure, the Pi's fan base isn't entirely academic. There exists a devoted following of do-it-yourselfers and hardware hackers who employ the Pi as an integral part of their hardware and software hacking experiments.

How about a solar-powered weather station? Or Pi-powered night vision goggles? How do you feel about having the ability to control your home's electronics from anywhere in the world using only your smartphone? All of these project ideas are eminently attainable at reasonable cost, thanks to the Raspberry Pi.

NOTE: WHAT'S IN A NAME?

In case you were wondering, the name *Raspberry Pi* does indeed have a colorful history. *Raspberry* pays homage to the fruit names that played a part in early-80s microcomputing: Apple Macintosh, Tangerine Microtan 65, Apricot PC—the list goes on.

Pi actually references not the standard number but the Python programming language. Eben and the rest of the Foundation originally thought that Python would be the sole programming language supported by their tiny personal computer. As you learn later, the RasPi allows enthusiasts to write programs using a large variety of programming languages.

Hardware Versions

It is important to remember that the Raspberry Pi is a full-fledged personal computer and not just a simple microcontroller. A *personal computer* is a self-contained system that performs the following data processing tasks:

- **Input**: The computer receives instructions and data from the user or application.
- **Processing**: The computer performs preprogrammed actions upon its input.
- **Output**: The computer displays the processing results in one or several ways to the user or application.

In addition, a personal computer typically also includes persistent storage and an operating system that features a user interface. Much more is discussed concerning these topics in Chapters 2 and 3, "A Tour of Raspberry Pi Peripheral Devices."

Suffice to say that the Raspberry Pi does essentially all the things that your full-sized desktop or laptop computer does, albeit more slowly and on a smaller scale.

By contrast, a microcontroller is a much more specialized piece of hardware. A *microcontroller* is an integrated circuit that is similar to a personal computer inasmuch as it receives input, performs processing on that input, and finally generates output of some kind or another.

However, the microcontroller is set apart from the personal computer by the following three characteristics:

- **A microcontroller's operation depends on precise timing**: Because the microcontroller is generally a single-purpose device, there's no driver or operating system overhead to slow down the system. Therefore, the microcontroller can perform work by using extremely precise clock cycles. This time-dependent operation is difficult to accomplish with the Pi because the Pi must access its hardware through several software layers.

- **A microcontroller gives the user full and direct access to hardware**: As you learn in Chapter 4, "Installing and Configuring an Operating System," most of the Raspberry Pi hardware (particularly the Broadcom BCM2835 system-on-a-chip) is proprietary. By contrast, most microcontrollers such as the Atmel Reduced Instruction Set Computing (RISC) chip at the heart of the Arduino are open source and are therefore completely accessible to users. With the Pi, we are limited to interacting with the board's hardware components via software application programming interfaces (APIs).

- **A microcontroller typically has no user interface**: A programmer must use an external system to send data to and receive data from a microcontroller.

- **A microcontroller is typically designed for a single purpose**: A microcontroller is intended to perform a single task—and to do that task precisely and very well. For instance, consider an Arduino-powered weather station that senses the environment and reports on air temperature, relative humidity, barometric pressure, and so forth.

A representative Arduino microcontroller (specifically the Uno) board is shown in Figure 1.2.

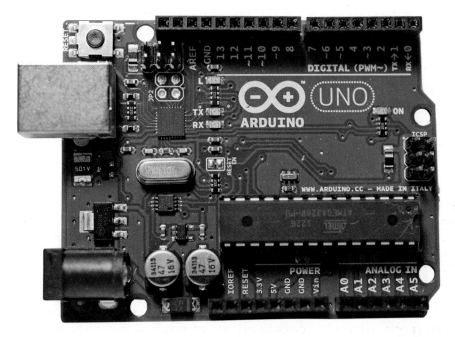

FIGURE 1.2 The Uno is the flagship of the Arduino microcontroller line.

You learn in Chapter 19, "Raspberry Pi and Arduino," how well the Raspberry Pi "plays" with the Arduino platform. In the meantime, here is a list that provides you with some of the most popular Raspberry Pi-compatible microcontrollers in use today:

- Arduino (http://arduino.cc)
- BeagleBone (http://beagleboard.org/bone/)
- Dwengo (http://www.dwengo.org/products/dwengo-board)

Now then, let's get down to business and formally introduce the Raspberry Pi.

As of spring 2013, the Raspberry Pi Foundation has two Raspberry Pi models, Model A and Model B. The differences between the two are shown in Table 1.1.

TABLE 1.1 Comparison of the Two Raspberry Pi Models

	Raspberry Pi Model A	Raspberry Pi Model B
Release Date	Q1 2013	Q4 2012
List Price	$25	$35
SoC	Broadcom BCM2835	Broadcom BCM2835
CPU	ARM11 700 MHz	ARM11 700 MHz
GPU	Broadcom VideoCore IV	Broadcom VideoCore IV
RAM	256 MB	512 MB
Power Rating	300 mA at 5V	700 mA at 5V
USB	1x USB 2.0	2x USB 2.0
Video Output	Composite RCA; HDMI 1.3/1.4	Composite RCA; HDMI 1.3/1.4
Audio Output	3.5mm analog; HDMI	3.5mm analog; HDMI
Networking	None	10/100 Mbps Ethernet

NOTE: HISTORY, HISTORY, ALL AROUND ME

Even the Raspberry Pi nomenclature pays tribute to the British Broadcasting Company (BBC) Micro personal computer. As it happens, the Micro BBC had a Model A and Model B, with Model B offering substantially more processing horsepower than the modest Model A.

So aside from the price difference, what are the key points of distinction between Model A and Model B? In the simplest terms:

- Model B has twice the RAM as Model A.
- Model B has an onboard Ethernet RJ-45 jack.
- Model B has an extra USB port.
- Model A uses 30% as much power as Model B.

If you spend time analyzing the two models (and I certainly hope you invest in one of each and do so), you'll notice that the printed circuit boards are indeed identical.

Instead of a redesign, the Foundation simply stripped components off Model B to make Model A. Look at the image of Model A in Figure 1.3, focusing on the lower right—see that empty socket? That's where the Ethernet port is soldered on the Model B board.

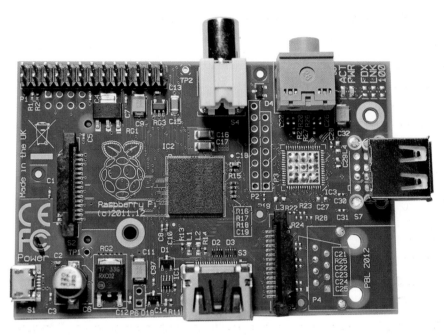

FIGURE 1.3   Raspberry Pi board, Model A.

Also notice in this image the unpopulated pad just above the Ethernet area; this is where the Foundation soldered an SMSC LAN9512 integrated circuit (IC) that controls the Ethernet jack in Model B.

Because the extra $10 buys you so much more computer, I focus on Model B exclusively in this book. The good news for Model A owners, however, is that all of the software, hardware, and programming we undertake here can be performed on both models.

NOTE: WHY MODEL A?

The question probably came to your mind, "Why would someone purchase Model A when you get so much more 'bang for your buck' with Model B?" The answer probably has something to do with power consumption. Because the power footprint of Model A is so tiny, Pi hackers can leverage the Model A platform for low-energy or solar-powered projects much easier than they can with Model B.

One more thing about the Model B boards specifically: As of spring 2013, the Raspberry Pi Foundation released two revisions to the Model B PCBs. You can tell at a glance which board revision you have in front of you by inspecting the light-emitting diode (LED) bank to the right of the USB port(s). Check it out:

- If the first status LED is labeled OK, you have a Revision 1 board.
- If the first status LED is labeled ACT, you have a Revision 2 board.

You learn more about the Pi's status LEDs in Chapter 2.

Is the Raspberry Pi Open Source?

A more important question to have answered before asking whether or not the Raspberry Pi is open source is "What does open source mean, and why should I care?"

Open source refers to hardware and/or software that is manufactured and given away free of charge with all intellectual property rights intact. For instance, the open source Linux operating system allows the general public to download, modify, improve, and release the underlying source code.

The term open source applies to hardware as well. For example, the schematics for the Arduino microcontrollers are freely available at the Arduino website (http://is.gd/VDVQfF); therefore, anybody in the world is allowed to analyze and understand the PCBs at a fundamental level.

Why do people invest time and money in developing open source hardware and software, only to release it to the public for free? Essentially, open source proponents are big fans of free information interchange.

Its open architecture is one important reason why Linux is considered to be one of the most secure operating systems in the world. When security vulnerabilities are identified, the Linux community can delve into the source code to identify and resolve the problem for the benefit of all Linux users across the world.

For comparison purposes, think of the Microsoft Windows and Apple OS X desktop operating systems. As you probably know, these OS platforms are proprietary, which means the general public cannot reverse-engineer the software to examine its underlying source code. The closed nature of proprietary software presents information security problems because only the software vendors themselves can resolve vulnerabilities that crop up in their products.

In conclusion, open source hardware and software offers increased security because the community can identify and correct vulnerabilities quickly. Open source architecture also lends itself to education because there are no proprietary, hidden components that bar learning. Finally, you can't beat the price of open source components—much of it is free, as previously discussed.

Open Source Licenses

Open source software is typically released under a license agreement called the GNU General Public License (http://is.gd/7s17wU), also called the GPL. The gist of the license is that anybody can download and use GPL software to their heart's content and for free. Users are also welcome to modify the software in any way that they see fit, as long as they release their modified version under the GPL.

NOTE: GNU WHO?

GNU, commonly pronounced *guh-NU*, is a recursive acronym that stands for "GNU's Not Unix." This is a super-geeky reference to the Unix operating system, the proprietary precursor to Linux. Incidentally, a gnu (pronounced *nu*) is a large, dark antelope that is also known as a wildebeest.

Open Source and the Raspberry Pi

The answer to the kernel question, "Is the Raspberry Pi open source?" is...well... complicated.

The Raspberry Pi runs variants of the Linux (also called GNU/Linux) operating system, which we've already established are free and open source. However, the "guts" of the RasPi's hardware—its Broadcom BCM2835 system on a chip (SoC)—is proprietary and closed source.

Remember that Eben and some other members of the Raspberry Pi Foundation have close ties to Broadcom. It's awfully nice of Broadcom to license its SoC for use in the Pi. The only downside, as I said, is that the intellectual property behind the Broadcom SoC is confidential.

If there is a silver lining to the Raspberry Pi closed source hardware situation, it is that in late 2012 Broadcom open sourced all of the ARM11 code for the Pi. What this means, especially with reference to the VideoCore graphics processing unit (GPU), is that the community can build their own device drivers that offer much more speed and optimization as compared to the default Broadcom drivers.

Upon closer inspection, however, the situation is much more complicated. The long story short is that Broadcom offers a GPL-licensed "shim" driver that cooperates input and output between the user and the CPU/GPU. For instance, the VideoCore IV driver itself consists of a proprietary, Broadcom-supplied binary large object (BLOB) driver that is not user-modifiable.

Just to be clear: The driver code for the Broadcom SoC is software and is at least somewhat open source. The hardware itself and its accompanying schematics, remain a mystery to all but Broadcom.

Why do Raspberry Pi enthusiasts care about this? Well, for starters, hardware hackers who want to access the full power of the VideoCore GPU need access to its source code to access its complete capability. Imagine if you bought a jigsaw puzzle that only gave you half of the pieces—would you feel somewhat limited in what you could do with that puzzle?

How Can I Purchase a Raspberry Pi?

The Raspberry Pi Foundation and its original equipment manufacturer (OEM) partners have worked hard to provide a supply chain for the Pi. According to Eben, the original plan was to manufacture Raspberry Pi boards strictly in the UK. However, scheduling and cost problems led the Foundation to initially seek Far East fabrication partners.

However, as of spring 2013, courtesy of the Sony factory in Wales, all Raspberry Pi production now occurs in the UK.

The finished units are sold exclusively by the following organizations:

- Premier Farnell/Element 14 (http://www.farnell.com/)
- RS Components (http://uk.rs-online.com/web/generalDisplay.html?id=raspberrypi)

Both Farnell and RS have distribution partners spread throughout the world; you shouldn't have trouble finding an official distribution source regardless of where you live.

The problem isn't so much finding a source for the Pi, but actually receiving a unit! Because demand has historically exceeded supply, during late 2012 and early 2013, I have observed backorders and long wait times from both Farnell and RS for both the Model B and the Model A boards. However, enough "critical mass" should develop in the Raspberry Pi supply chain that by the time you read this, availability should be reasonable.

One alternative you might want to consider is purchasing your Pi from a reputable eBay seller. I myself have had great luck in that regard. You know the rules with supply and demand—you'll typically pay a premium over the $35 list price for a Model B board. On the other hand, you can get productive with the Pi months before you can receive one from Farnell or RS.

The bottom line, friends, is that if you want to get your hands on a Raspberry Pi, you'll most likely need to do so by purchasing the device from an online reseller. And then perhaps the day will come (hopefully soon) when consumer electronics stores such as RadioShack and Best Buy stock these fascinating devices.

## CHAPTER 2

Hardware Components Quick Start

Have you ever heard the term "black box"? *Black box* is meant to denote any object or process whose inner workings remain outside of easy view. Personal computers, for instance, are often referred to as black boxes. Thus, some people are afraid of troubleshooting their own computers for that very reason—just what *does* lie inside that fancy computer case?

Well, if you've received your Raspberry Pi, you might have been surprised to see that the Foundation doesn't give you a case (or peripherals, for that matter). That's right—for $35 you receive the Raspberry Pi Model B board and nothing else.

This is actually good news because it forces us to take a long, hard look at that befuddling mass of wires, components, solder joints, and doo-dads that reside on the Pi board.

By the time you finish this chapter, you won't be intimidated by the raw Raspberry Pi anymore. At the least, your decision to purchase an aftermarket case for the thing will be driven by practical concerns rather than any other reason.

Shall we begin?

Understanding Pi Hardware Terminology

Both Model A and Model B of the Raspberry Pi are what are called *printed circuit boards*, or PCBs. A PCB is a laminate sheet that provides a connection platform for one or more electrical circuits.

A PCB ordinarily consists of some or all the following components, all of which are surface-mounted to the board:

- **Traces**: These are the tiny copper wires that are embedded into the PCB substrate and that form the backbone of the circuit(s).

- **Vias** (pronounced *VEE-ahs*): These are small metal rings that serve as an interconnect for boards that have multiple layers of traces.

- **Resistors**: These are components that resist the flow of electrical current. Resistors are labeled R*X* on the PCB, with *X* denoting a discrete number identifying a specific capacitor.

- **Capacitors**: These are components that temporarily store electrical charge. Capacitors are labeled C*X* on the PCB.

- **Diodes**: These are components that force electrical current to flow in a particular direction. Diodes are labeled D*X* on the PCB.

- **Transistors**: These are three-terminal components that act as electrically controlled switches.

- **Integrated Circuits (ICs)**: These are self-contained modules that run a circuit of their own in addition to interacting with circuits on the PCB. The Broadcom SoC is itself an integrated circuit module.

Take a look at Figure 2.1, which shows the back (bottom) side of Model B.

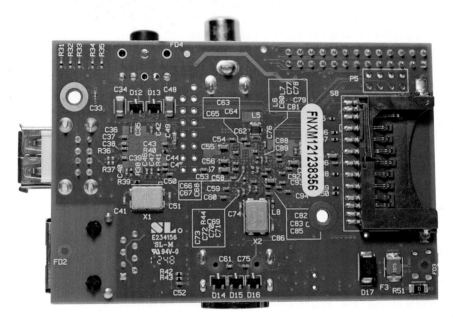

FIGURE 2.1   The back side of the Raspberry Pi Model B board.

That's quite a web of traces and solder joints, wouldn't you agree?

Recall from Chapter 1, "What Is the Raspberry Pi," that the heart and soul of the Pi is the system on a chip (SoC, pronounced *sock*) called the Broadcom BCM2835. Actually, the SoC is an IC that is sandwiched between the PCB (below) and the RAM memory chip (above). You can see the SoC/RAM chip duo in the dead center of the Raspberry Pi PCB; see Figure 2.2 for details.

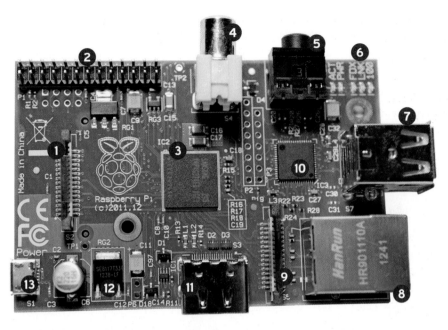

FIGURE 2.2   The front side of the Raspberry Pi Model B board.

1. DSI video

2. GPIO

3. CPU/GPU/RAM

4. RCA video

5. Stereo audio

6. Status LEDs

7. USB

8. Ethernet

9. Camera

10. USB/Ethernet controller

11. HDMI

12. Voltage regulator

13. Power Micro USB

The Foundation decided to stack the memory IC on top of the Broadcom SoC to save space on the board. That it does, I must say.

The processor is the mathematical muscle of any computer system. The Pi features an ARM11 processor that runs at a base speed of 700 megahertz (MHz). This means that the Pi (at least in theory) can execute 700 million instructions per second, although some instructions require a couple clock cycles.

Specifically, the ARM chip employs an instruction set called ARMv6. Some advanced hardware geeks complain that the Pi's processor does not support the ARMv7 architecture that is featured in other ARM-equipped mobile devices such as the BeagleBone, but in my opinion, it has all the computing power needed for general-purpose enthusiasts.

Ultimately, the fact remains that the main day-to-day difference between ARMv6 and ARMv7 is that the latter is faster even at the same processor clock speed. We are also faced with the fact that the Raspberry Pi is somewhat more limited in the software that it runs due to the ARM version.

The ARM processor is used most widely in mobile devices; it is actually sort of novel for a personal computer such as the Raspberry Pi to run this chip. Unfortunately, the ARM architecture means that the Pi cannot run Windows or Mac programs (that is, on those systems that operate on the Intel/AMD 32- and 64-bit processor platforms).

Another point to consider about the Broadcom BCM2835 is that the SoC delivers not only the ARM central processing core, but also the onboard video processor—remember that SoC stands for *system on a chip*. As it happens, the BCM2835 includes the VideoCore IV GPU (pronounced *gee-pee-you*), which stands for graphics processing unit.

So, in other words, the SoC contains two "brains"—the main ARM processor for general-purpose number crunching and the VideoCore IV GPU for video graphics display. What's especially cool about the VideoCore IV is that it can decode and play full 1080p high definition (HD) video by using the vendor-neutral industry standard H.264 codec.

You learn in Chapter 18, "Raspberry Pi Overclocking," how you can adjust the memory split between system memory and graphics memory. Doing so optimizes the Pi for particular types of applications and uses.

The Hidden Cost of Owning a Raspberry Pi

As you already know, your $25 or $35 gets you a Raspberry Pi PCB and nothing else. This comes as a surprise to many people, so I want to give you all the prerequisites as early in the game as possible.

At the very least, you'll need to purchase or otherwise obtain the following hardware to get started with your Raspberry Pi:

- Micro USB Power Supply
- SD Card
- Powered USB hub

- Ethernet cable
- Monitor
- Video cable
- Audio cable
- USB keyboard and mouse

The keyboard, mouse, and monitor/cable are technically optional if you plan to run the Raspberry Pi "headless" (headless Pi setup is covered in Chapter 7, "Networking Raspberry Pi").

Now let's explain each of these required hardware components in greater detail.

A 5V Power Supply

The Raspberry Pi expects an incoming electrical voltage of 5 volts (V) ±5% per the USB 2.0 standard. I mention the Universal Serial Bus (USB) here because power to the Pi is derived from a USB-based power supply. From there the Pi's electrical requirements depend on which board you have:

- **Model B**: 700 milliamps (mA) at 3.5 watts (W) or
- **Model A**: 500 mA at 2.5 W

NOTE: A REAL POWER SAVER

The dramatically less power that Model A requires represents the board's chief attraction among electronics enthusiasts. We can build super low-power projects with Model A that would be nearly impossible to do with Model B. The reason for Model A's lighter power consumption footprint has to do principally with the absence of the SMSC LAN9512 Ethernet controller, which pulls quite a bit of power on its own. On the other hand, remember that if you plug a USB Wi-Fi dongle directly into the Pi, power consumption goes up correspondingly.

The power port on the Raspberry Pi PCB is the Micro USB B-style interface; therefore, a Pi-compatible power supply uses the standard USB A connector on one side and the Micro USB B connector on the other side. I show you my own Raspberry Pi power supply in Figure 2.3.

FIGURE 2.3 A Raspberry Pi-compatible power supply on the left (Micro USB plug is labeled 1) and a powered USB hub on the right (Standard USB plug is labeled 2).

It is crucial for you to understand that although the Raspberry Pi expects 5V of power incoming to the board, the board internally operates at a much lower 3.3V. This power step-down is accomplished automatically by virtue of the Pi's on-board voltage regulator and C6 capacitor.

Not to get too geeky with the electronics (I'll work up to that stuff gradually throughout this book), a *capacitor* is an electronics component that stores an electrical charge temporarily. If you examine the barrel-shaped object in Figure 2.2 just to the rear of the Micro USB interface, you'll see the C6 capacitor. By the way, C6 refers to the PCB label for this component.

The C6 capacitor is really quite cool. It ensures that the incoming 5V is smooth and steady. If you have a cheap power supply that dips or spikes every so often, the capacitor can step in and even out the voltage flow. Pretty neat!

When the Pi's circuitry has been expanded to, say, a microcontroller board or a breadboard, you need to keep voltage regulation at the forefront of your mind to avoid overpowering the Pi and causing irreversible damage.

In Chapter 4, "Installing and Configuring an Operating System," I recommend that you stick to known name brands when you choose a Secure Digital (SD) card for your Pi's operating system. You'll want to follow this rule when choosing a power supply as well.

Some of my colleagues and friends have been burned by purchasing off-brand "Raspberry Pi-compatible" power supplies from eBay or Craigslist vendors. The main problem with cheap power supplies is that they don't deliver a solid 5V of direct current (DC) to the PCB. If the power supply delivers too much juice, the board can fry. If the power supply doesn't deliver enough power, the Pi will shut down at worst and operate erratically at best.

The following are a few third-party vendors who do produce reliable Raspberry Pi-compatible power supplies:

- AdaFruit (http://is.gd/klOOWr)
- ModMyPi (http://is.gd/Tme3iq)
- SparkFun (http://is.gd/rs6UJx)

SD Card

The Secure Digital (SD) card is a solid-state removable storage device that is needed because the Pi has no permanent, onboard data storage capability.

You learn how to flash Raspberry Pi firmware and a Linux operating system to the SD card in Chapter 4. For now, however, just know when you start shopping that you are looking for the following items:

- A standard SD card. (The SD specification consists of the Standard, Mini, and Micro form factors. You can use an adapter to convert a Mini or a Micro SD card into a Standard size.)
- A brand name product, rather than cheaper, generic options. Some of the SD card brands I trust include Kingston (http://is.gd/j9kb1O), Transcend (http://is.gd/Jwxe1N), and SanDisk (http://is.gd/2NZw8b).
- Capacity of at least 4 GB
- Class 4 or higher

The speed class rating of an SD card is a relative indicator of how quickly the card can read and write data. Do you remember the old CD burners with their 2x, 4x, 48x nomenclature? Same thing here. A Class 2 spec exists, but I would stick with one of the following SD speed classes:

- **Class 4**: 4MB/sec
- **Class 6**: 6MB/sec
- **Class 10**: 10MB/sec

You'll find that opinions vary widely regarding which SD card brand(s) or speed(s) is optimal. My best advice to you is to try out a few models and speed ratings with your equipment and let your own intuition and Pi benchmark results be your guide.

Standard-sized SD cards come in two varieties: Secure Digital High Capacity (SDHC) and Secure Digital eXtended Capacity (SDXC). The main difference is that SDHC cards go up to 32GB, and SDXC cards go up to 2TB. If I were you, I would check the Raspberry Pi compatibility list at http://is.gd/Ym6on0 before I shell out the money for a top-of-the-line, highest-capacity, speediest card. Sometimes it is best to go for compatibility instead of potential performance.

Powered USB Hub

The Model B board includes two USB ports, but please don't let that "security" lull you away from the reality that you truly need to purchase a powered USB hub. Some Raspberry Pi newcomers use the two USB ports for keyboard and mouse connections and then scratch their heads in wonderment when they realize, "How the heck can I plug in something else to the Pi?"

A hub is a compact device that hosts several USB A-type devices. The "powered" part is important inasmuch as USB hardware has in itself a current draw. Thus, we need to ensure that our USB hub can supply not only the 700 mA required by the Pi board, but also any power requirements for USB-attached peripherals.

Actually, that point bears repeating: Ensure that any power supply that you consider for your Pi supplies at least (but hopefully more) than 700 mA. Non-self-powered USB peripherals will each draw 100 mA or so from the USB ports on your Pi. To be sure, you should consider a powered USB hub as a "must have" peripheral for your Pi.

Something else: The power supply and a powered USB hub are two separate pieces of hardware and serve different purposes. The USB power supply gives power to the Raspberry Pi itself and allows it to function. A powered USB hub enables you to expand the Pi's functionality by adding more hardware and giving power to those additional devices rather than to the Pi.

You can see what my own powered USB hub looks like by examining Figure 2.3.

Ethernet Cable

If you want to connect your Pi to the Internet (and why wouldn't you want to do that?), you'll need an Internet connection and a Category 5e or 6 Ethernet cable. The Model B board includes an onboard RJ-45 Ethernet jack, into which you plug your new cable. You plug the other end of the cable into a free switch port on your wireless router, cable modem, or Internet connectivity device.

NOTE: CONNECTORS AND PORTS

In physical computing, a *port* or *jack* is the connection interface on the computer. The perimeter of the Raspberry Pi, for example, is lined with ports of different varieties.

A *plug* or *connector* is the part of a cable that plugs into a port. For instance, a Category 6 Ethernet cable uses an RJ-45 connector to plug into the RJ-45 jack on the edge of the Raspberry Pi Model B board.

"But what about Wi-Fi?" you ask. Wi-Fi and all other network-related questions are addressed in Chapter 7. For now, understand that if you have a Model A board, your only option for traditional wired Ethernet networking is to purchase a USB wired Ethernet adapter. Again, more on that subject later on in the book.

The subject of Wi-Fi connectivity bears on what we just covered vis-a-vis USB ports and powered USB hubs. That is to say, we must use a USB Wi-Fi dongle in order to give wireless Ethernet connectivity to our Raspberry Pi device.

Monitor

Unless you plan to run your Raspberry Pi remotely in a so-called headless configuration, you need to set aside a spare monitor or television for use with your Pi. Yes, you heard me correctly: You can plug your Pi into any television, be it an older model (via an old-school yellow RCA plug) or a modern HD display using the HDMI interface.

In fact, one of the Raspberry Pi Foundation's goals in designing the Pi was to support "any old" television set as a cheap display device. Remember that the Foundation's philosophy is to make the Raspberry Pi as inexpensive and easy as possible for people to get their hands on and to start programming.

NOTE: TINY LITTLE SCREENS

Some Raspberry Pi enthusiasts translate the tiny footprint of the Pi board into the monitor as well. To that point, you can purchase small (think GPS-sized) color monitors from a number of online retailers. For instance, check out this adorable 7-inch diagonal HDMI display from AdaFruit: http://is.gd/GJARAZ

Cables

Depending on what type of monitor or TV you have at your disposal, you might need to purchase an analog RCA video cable or a digital HDMI cable. The good news is that these cables are almost ubiquitous and are quite inexpensive.

As you learn in more detail momentarily, the use of an HDMI cable means that you don't have to worry about providing audio-out capability in your Pi with an analog audio cable. However, if you're using the RCA video cable and do need audio, you'll need to buy a 3.5mm stereo audio cable as well.

NOTE: HDMI WITH DEDICATED AUDIO

In case you were wondering, it is possible to configure the Raspberry Pi to use the HDMI cable for video and the 3.5mm stereo audio cable for audio. To do so, you must instruct the Pi to disable the HDMI audio channels by running the command **sudo amixer cset numid=3 1** from a Raspbian shell prompt. By the way, Raspbian is the official Linux distribution of the Raspberry Pi; we'll learn all about it beginning in Chapter 4.

USB Keyboard and Mouse

The good news is that the power draw for USB keyboards and mice is low enough that you can plug them directly into the USB interfaces on the Model B board. The bad news is that you won't have any additional expandability for your Pi. Therefore, your best bet is either to invest in the previously described powered USB hub or connect to your Pi remotely.

Figure 2.4 shows a Raspberry Pi all plugged in.

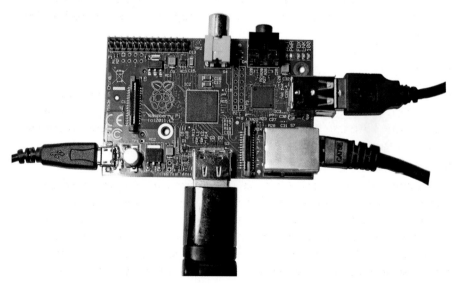

FIGURE 2.4   A Raspberry Pi, fully connected and ready to go!

CAUTION: JUST IN CASE

As you can see in Figure 2.4, a "naked" Raspberry Pi, especially when it's all cabled up, is quite vulnerable to your physical environment; this includes electrostatic discharge (ESD) as well as physical factors. For these reasons you should consider purchasing a case for your Pi. You read about cases in more detail in Chapter 3, "A Tour of Raspberry Pi Peripheral Devices." Even with a case, however, you should take steps to avoid ESD when interacting with the Pi hardware. You can do this by using an antistatic wrist strap whenever you handle the bare Pi board.

A Tour of the Model B Board

Now, let's commence our tour of the Model B board. The tour begins with the lower-right of the board from the perspective in Figure 2.2.

Networking

The cube-shaped module on Model B is the onboard *Registered Jack 45 (RJ-45) Ethernet interface*. As some of you might know, wired Ethernet is capable of running at data transmission speeds of 10, 100, and even 1000 megabits per second (Mbps).

However, because the Raspberry Pi Ethernet interface operates using the USB 2.0 standard, the jack is limited to either 10 or 100 Mbps. As long as you purchase a Category 5e or 6 Ethernet cable and your network already operates at 100 Mbps, the Pi should work at that speed with no problem at all.

Video and Audio

In my opinion, the preferred way to handle outgoing video and audio is to employ the Pi's integrated *High Definition Multimedia Interface (HDMI) port*. The number one reason is that HDMI carries both video and audio signals. And number two, the signaling is entirely digital. HDMI is the way to go if you plan to use your Raspberry Pi as a multimedia center because you have access to full HD 1920x1080 screen resolution.

The only possible downside to using HDMI is that only later-model computer monitors support the interface. Many LCD monitors still in wide use only support DVI connectors. In this case, you still have a couple options: First, you can plug your Pi into your HDMI-compatible television; second, you can buy an HDMI-to-DVI-D converter plug and connect the Pi to your Digital Video Interface (DVI-D) -equipped computer monitor; this is shown in Figure 2.5. Of course, if you do this you lose the ability to carry audio as well as video.

FIGURE 2.5   You can easily convert an HDMI connection to a DVI-D connection.

NOTE: VGA NEED NOT APPLY

The 15-pin Video Graphics Array (VGA) port that is found in older monitors is incompatible with the Raspberry Pi.

The yellow circular jack opposite to the HDMI port on the board is what is called the *RCA connector* and forms the video feed portion of an old multimedia standard called composite video. Although this plug allows you to connect your RasPi to ancient television sets, the signaling is analog, the video capability is standard-definition only, and there is no signaling left over for audio.

Model B includes a third video interface just to the left of the Raspberry Pi logo called the *Display Serial Interface, or DSI*. This display interface is used primarily for tablet or smartphone touch screens (remember that the ARM architecture in general is slanted heavily toward the smartphone market). As of spring 2013, little is published on how to make use of the DSI interface. Recall that we don't have full access to the VideoCore IV GPU, so we mere mortals cannot yet develop a kernel-mode driver for this interface. Keep your eyes peeled online because I'm sure we'll see development in this area before too long.

As an alternative to HDMI audio, the Raspberry Pi includes a *3.5 mm stereo audio jack*. This means you can connect computer speakers or perhaps headphones to your Pi to receive analog audio from the board.

Storage

Many Raspberry Pi newcomers are befuddled as to where to connect their SD cards to the board. You'll find that the *SD card slot* is a bare-bones port that is actually mounted underneath the PCB. Thus, you line up the SD card on the interface rails and gently push the card until it is fully seated on the interface pins.

Don't worry that the SD card sticks out from the side of the Pi a little bit—that behavior is by design to facilitate card removal. Actually, the fact that SD card pokes out from beneath the PCB is yet another reason for you to invest in a Raspberry Pi case.

With respect to volatile (nonpermanent) storage, don't forget about the 512MB random access memory (RAM) chip that is stuck directly on top of the SoC at the center of the PCB. Recall also that the Model A board includes a 256MB RAM chip.

Power/Status Information

The *Micro USB power port* intends to supply 5V of direct current (DC) to the board from your external power supply. Recall, however, that the Raspberry Pi operates at an internal voltage of 3.3V. The good news is that the Pi board includes an onboard voltage regulator (located behind the Micro USB port in the location marked RG2), as well as the C2 capacitor to smooth out the voltage.

In one corner of the Model A or Model B board, next to the stereo audio jack, you'll observe a bank of *light emitting diodes*, or *LEDs* (see Figure 2.6). These LEDs light up to denote the following status conditions:

- **ACT (Green)**: SD Card Access
- **PWR (Red)**: 3.3 V power present
- **FDX (Green)**: Full Duplex LAN connected
- **LNK (Green)**: LAN link activity
- **100 (Yellow)**: 100Mbps LAN connected

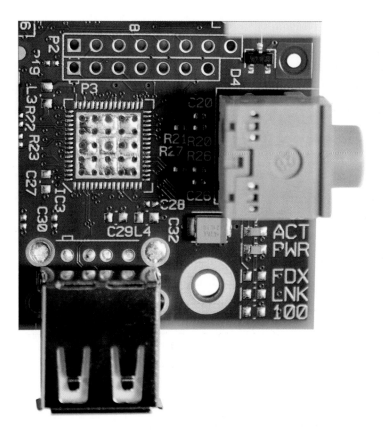

FIGURE 2.6 LEDs give you at-a-glance status information from your Raspberry Pi. In this image, notice the bank of LEDs in the lower right corner of the PCB.

Remember that you can tell at a glance whether you have a Revision 1 or Revision 2 board by examining the label of the first LED. If the LED is labeled ACT, you have a Revision 2 board. If the label reads OK, you have a Revision 1 board.

Camera

The Model B board includes a Mobile Industry Processor Interface (MIPI) *Camera Serial Interface (CSI)* connector; the interface is located just behind the Ethernet port and connects to the *Raspberry Pi camera board* that the Foundation released in May 2013.

The heart of the Raspberry Pi camera module is a 5 megapixel (MP) Omnivision OV5647 sensor that shoots still images at a 2592x1944 pixel resolution and records 1080p/30 frames-per-second video.

The Foundation sells the camera board through its usual distribution partners for $25. The use of the camera module is covered in great detail in Chapter 16, "Raspberry Pi Portable Webcam."

Processing

As previously discussed, at the center of the Raspberry Pi Model B board is a two-layer integrated circuit (IC) stack called a *chipset*. On top is a 512MB random access memory (RAM) module. At the bottom is the Broadcom BCM 2835 SoC.

Remember that the SoC consists of two processor cores: a 700 MHz central processing unit (CPU) that is used for general computing tasks and a VideoCore IV graphics processing unit (GPU) that is used for, well, video generation.

Expansion

Okay—I've saved the best for last. On the same side of the board as the status LEDs but on the opposite end is a bank of 26 copper header pins called the *General Purpose Input/Output (GPIO) interface*.

The GPIO is critically important to the Raspberry Pi because these pins represent the way we can expand the Pi board to interact with external hardware such as microcontrollers, motors, robotics—you name it!

You learn the specific purpose of each GPIO pin—called, appropriately enough, the pinout—in Chapter 19. In the meantime Figure 2.7 provides you with an illustration of how you can use the GPIO header.

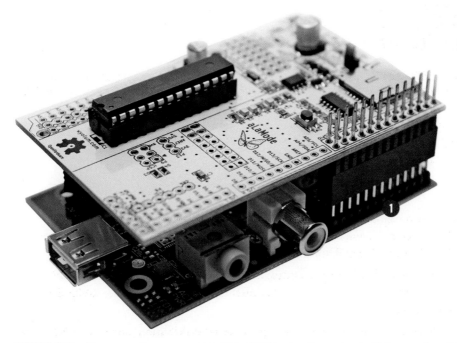

FIGURE 2.7 You can leverage the Pi's GPIO pins (marked 1) to work with expansion boards such as the Alamode (http://is.gd/6eMMnC). The Alamode is an Arduino clone that can broaden and deepen the capabilities of the Raspberry Pi.

It seems the subject of Raspberry Pi cases has arisen several times in this chapter. I advise you to be choosy when selecting a case for your Pi. Some cases look cool but actually can heat up your Pi board due to insufficient venting. Moreover, I've seen some Pi cases that make it difficult or impossible for you to access the GPIO pins with the case in place.

I have had good luck with Raspberry Pi cases purchased from Adafruit (http://is.gd/K1Ow9s), the Pi Hut (http://is.gd/hR22tW), and ModMyPi (http://is.gd/xT8LTA).

Next Steps

I hope you are now more comfortable with the Raspberry Pi hardware. Now that you understand how the Raspberry Pi board is set up, you probably want to know more about the extra hardware that can be plugged into your Pi to expand its capabilities. To that point, let's dive into a detailed consideration of Raspberry Pi peripheral devices.

## CHAPTER 3

A Tour of Raspberry Pi Peripheral Devices

I got my start learning about electricity and electronics not through school but by horsing around with a Science Fair 160-in-1 electronics project kit my parents bought for me from Radio Shack for my tenth birthday.

As you can see in Figure 3.1, this wooden-framed kit enabled kids like me to prototype electrical circuits without having to solder any components together. The various "doo dads" on the kit's circuit board kept me engaged and entertained for many, many hours.

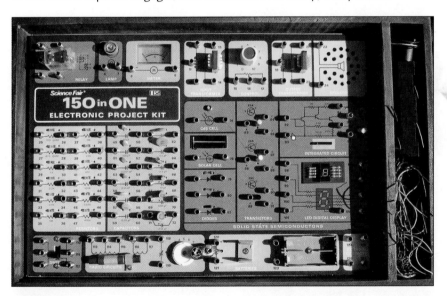

FIGURE 3.1 I learned electronics by studying (playing?) with this Radio Shack project kit.

Fast-forward to the twenty-first century—now we have the Raspberry Pi, a $35 personal computer the size of a credit card! In this chapter, I'd like to pique your curiosity by sharing with you the most popular peripheral devices—which is to say, electronic equipment that is connected to the Pi by means of a cable instead of soldered directly to the board—that exist in today's marketplace.

If you want to really dig into physical computing and circuit building, you will indeed need to take an iron and braid in hand and learn to solder. I have you covered, though: You learn about all of the most popular starter kits and technician tools at the end of this chapter.

Let's begin!

Circuit Prototyping Equipment

In electronics, *prototyping* refers to mocking up an idea in a way that the circuit can easily be rebuilt. To that end, the breadboard is by far one of the most useful tools you can have in your possession.

A *breadboard* is a plastic block that is perforated with small holes that are connected internally by tin, bronze, or nickel alloy spring clips. Take a look at Figure 3.2 as a reference while I explain how these devices work.

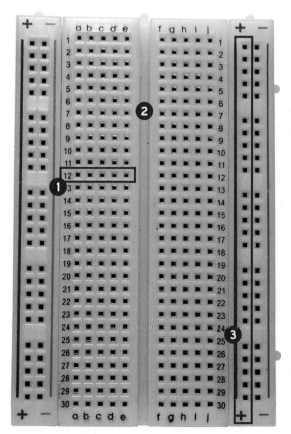

FIGURE 3.2 Anatomy and physiology of a breadboard. A terminal strip is labeled 1, the bridge is labeled 2, and a bus strip is labeled 3.

First of all, see the empty area that runs down the center line of the breadboard? This region is called the *bridge*. It's a physical barrier that prevents current on one side from interacting with current on the other side. Thus, the breadboard is *bilaterally symmetric*, which is a fancy way of saying it consists of two mirror image halves that represent two separate circuits.

When you mount integrated circuit chips that use the dual inline package (DIP) format on a breadboard, be careful to align the opposing sets of pins on opposite sides of the bridge to prevent circuit overflow.

If you are wondering what a DIP looks like, whip out your Raspberry Pi board and look below the GPIO header: the voltage regulators labeled RG1, RG2, and RG3 are DIPs.

The horizontally numbered rows of perforations represent the breadboard's *terminal strips*. Any wires that you connect in a single row share a single electrical circuit. Breadboards come in several different sizes, and each has its own number of terminal strips.

For instance, full-sized breadboards typically include 56 to 65 connector rows, while smaller breadboards normally have 30 rows.

Finally, there are the horizontally aligned perforations that line the outer edges of the breadboard. These are called *bus strips*, and they constitute "power rails" for your prototype circuits. One connector column represents supply voltage (positive), and the other represents ground (negative).

In sum, the breadboard is the perfect platform for prototyping electrical circuits because you don't need to solder anything. Instead, you can simply "plug and play" with ICs, resistors, lead wires, buttons, and other components.

Of course, all of this background information on breadboarding suggests the question, "Why would I, a Raspberry Pi owner, want to prototype anything?"

Great question! Here's the deal: If you want to use your Raspberry Pi to interact with the outside world, whether that interaction is controlling a robot, snapping pictures from 30,000 ft in the air, or creating a solar-powered weather station, you'll need to learn how to use prototyping hardware such as breadboards, resistors, jumpers, and the like.

On the Raspberry Pi, the 26 General Purpose Input/Output (GPIO) pins are used to "break out" the Pi onto a breadboard. You can do this by using two different types of cable:

- **Ribbon cable**: This flat cable connects to all the GPIO pins simultaneously
- **Jumper wire**: This wire connects a single GPIO pin to a terminal on the breadboard. Jumper wires are also called straps, and you'll use several of them when we use the Gertboard expansion board in Chapter 20, "Raspberry Pi and the Gertboard."

A ribbon cable and jumper wires are shown in Figure 3.3.

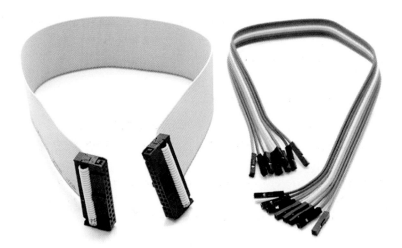

FIGURE 3.3 Ribbon cable at left and female-to-female jumper wires at right.

Breakout boards provide an excellent and convenient way to connect your Raspberry Pi to a solderless breadboard. I recommend the Pi Cobbler kit, sold by Adafruit Industries (http://is.gd/b4LIQ7).

As you can see in Figure 3.4, you mount the Pi Cobbler board across the breadboard bridge (do you like my alliteration?). The ribbon cable connects from the Cobbler to the Pi's GPIO header on the other side of the connection.

FIGURE 3.4 The Pi Cobbler is a quick and easy way to expand your Raspberry Pi to a breadboard.

Once you've broken out your Pi to the breadboard, you have the proverbial world available to you. In point of fact, the latter part of this book walks you through some real-world projects that take advantage of the Raspberry Pi-breadboard connection.

Single-Board Microcontrollers

Recall from our initial discussion in Chapter 1, "What is the Raspberry Pi?" that a microcontroller is a PCB that is designed primarily for a small number of time-dependent tasks.

The big benefit of integrating your Raspberry Pi with a microcontroller is that you can connect to an almost endless number of analog and digital sensors. This means you can write programs that detect and take action on the following and more:

- Light
- Moisture
- Sound/Volume
- Contact
- Motion

The Arduino platform (www.arduino.cc) is a suite of electronics prototyping PCBs that are dearly loved by artists, designers, inventors, and hobbyists for their ease of use and flexibility.

Hobbyists have developed some pretty cool technology by using Arduino microcontrollers: motion sensors, home automation systems, MIDI controllers, radon detectors...the list of project ideas is seemingly endless.

The Raspberry Pi–Arduino heavenly match is discussed in Chapter 19, "Raspberry Pi and Arduino." For now, however, let's go over the basic "gotchas" of this electronic marriage:

- **Connection options**: To connect your Raspberry Pi to an Arduino board, you can either use a USB cable or a I²C (pronounced *eye-squared-see*) serial link. You can see the Pi and Arduino UNO lined up side-by-side in Figure 3.5.

- **Voltage differences**: We already know from Chapter 2, "Hardware Components Quick Start," that the Raspberry Pi accepts 5V inbound power but operates at 3.3V internally. By contrast, the Arduino operates externally and internally at 5V. Consequently, when joining Pi with Arduino you need to invest in an external voltage regulation solution to avoid burning up your Pi.

FIGURE 3.5 You can connect an Arduino board directly to the Raspberry Pi by using USB, serial, or GPIO connections.

- **Administration**: Recall that the lack of an operating system is one of the defining characteristics of a single-board microcontroller. Therefore, in an Arduino/Raspberry Pi nexus, all your Arduino programming happens on the Pi, and you upload your Arduino "sketches" to that hardware over the serial or USB connection.

NOTE: BUT WAIT, THERE'S MORE!

Perhaps the most elegant way to connect your Arduino board to your Raspberry Pi is to purchase the Alamode shield (http://is.gd/4H3aWv). The Alamode is an Arduino device that connects directly to the Pi's GPIO header and provides a real-time clock, seamless connectivity to the Arduino microcontroller application programming interface (API), and voltage regulation to the Pi. It's a great deal!

Please note that despite its overwhelming popularity, the Arduino is not the only single-board microcontroller game in town. Here's a quick list of single-board microcontroller vendors that you might find useful:

- Texas Instruments MSP430 LaunchPad (http://is.gd/xbAjcO)
- Teensy (http://is.gd/rlElxy)
- STM32 (http://is.gd/TscRtp)
- Pinguino (http://is.gd/rdEpF5)

The Gertboard

What the heck is a Gertboard, you ask? Gert van Loo is a computer electronics engineer who was the chief architect of the Raspberry Pi PCB. Gert designed the Gertboard as a Raspberry Pi expansion board, or *daughterboard*, that makes it easy to detect and respond to physical (analog) events such as voltage changes, motor state changes, and the like.

NOTE: WHAT'S A DAUGHTERBOARD?

A *daughterboard* is a printed circuit board that is intended to extend the functionality of mainboard. In this context, the Raspberry Pi is the mainboard, and the Gertboard is the daughterboard. In full-sized PCs, daughterboards, which are also called mezzanine boards or piggyback boards, are often used to enable expansion cards to mount on their side, parallel to the motherboard, in the name of making the PC's form factor as slim as possible. Question: Do you think the world needs a brotherboard? How about a second-cousin-twice-removedboard?

The Gertboard is awesome because it saves you the work of building circuits with a breadboard. The Gertboard PCB is literally covered with useful electrical components like the following:

- Tactile buttons
- LEDs
- Motor controllers
- Digital-to-analog and analog-to-digital converters

The Gertboard and the Pi connect together directly by means of (what else?) the Pi's GPIO header. Figure 3.6 shows you a close-up of the amazing Gertboard.

FIGURE 3.6 The Gertboard provides a truly seamless expansion experience for the Raspberry Pi.

Just wait—you get to use the Gertboard in Chapter 20.

Single-Board Computers

The BeagleBone (http://is.gd/A5m89F) is perhaps the Raspberry Pi's chief competitor in the single-board computer market. The BeagleBone is, like the Raspberry Pi, an ARM-based, credit card-sized Linux computer.

The BeagleBone is actually the smaller sibling of the BeagleBoard. Both boards are manufactured by the legendary Texas Instruments (TI), which lends immediate credibility to the Beagle projects.

Serious gearheads prefer the BeagleBone because its ARM Cortex A8 processor (running at 720 MHz) supports the ARMv7 instruction set, as opposed to the ARMv6 set included with the Pi.

Because of its support for ARMv7, the BeagleBone's benchmark performance is much better than that of the ARMv6-equipped Raspberry Pi. You also have a wider range of Linux distributions to choose from with ARMv7-compatible devices such as the 'Bone.

The BeagleBone and the Raspberry Pi aren't exactly "finger in glove" partners like the Arduino and Pi. Really, they are competitors in the same or highly similar market space. Table 3.1 compares and contrasts the technical specifications for both systems.

TABLE 3.1   Comparison Between the Raspberry Pi Model B and the BeagleBone

	Raspberry Pi Model B	**BeagleBone**
List Price	$35 USD	$89 USD
Dimensions	3.37" x 2.125"	3.4" x 2.1"
SoC	Broadcom BCM2835	TI AM3359
CPU	700 MHz, single core	720 MHz, single core
GPU	VideoCore IV	PowerVR
RAM	512MB	256MB
USB	2	2
Storage	SD	MicroSD
GPIO	Yes	Yes
Camera	Yes (future development)	No
Development	Squeak/Scratch; Python	Squeak/Scratch; Python
Video Out	HDMI; Composite	None

Here's a noncomprehensive list of other single-board computer manufacturers:

- Cotton Candy (http://is.gd/quXmJu)
- CuBox (http://is.gd/B6hvsZ)
- Gumstix (http://is.gd/29EJ4A)
- PandaBoard (http://is.gd/5of9yx)

"Why is it important that I understand the Raspberry Pi's competition?" you might ask. In my estimation, it is important for you to know that there exist alternatives to the Raspberry Pi. You may find, for example, that the Pi is the best fit for the types of learning goals and projects that you have in mind. By contrast, you may also save yourself time, money, and frustration by concluding at the outset that you should consider an Arduino or a BeagleBone rather than a Pi.

Relevant Technician Tools

To perform all aspects of physical computing with your Raspberry Pi, you'll need a few electrical tools. Chief among these is the *digital multimeter*, an instrument with which you can measure electrical current, voltage, and resistance.

For instance, you can run a quick verification of the Raspberry Pi's 5V power supply voltage by using a multimeter and the TP1 and TP2 test points on the Model B board.

To locate the TP1 and TP2 test points, take a closer look at Figure 2.2. You'll see TP1 located just above and to the right of the C2 capacitor and TP2 just to the left of the RCA video output (assuming you are looking at the PCB with the Raspberry Pi logo facing up).

TASK: CHECK RASPBERRY PI VOLTAGE WITH A MULTIMETER

1. Turn on your multimeter and set the dial to a low voltage threshold (for instance, 20V is good). If you own an autoranging multimeter, you don't need to worry about this step.

2. Disconnect all peripherals (including the SD card) from your Pi except for the Micro USB power supply cable.

3. Power up your Pi.

4. Place the hot (red) lead to the TP1 test point, and place the ground (black) lead to the TP2 test point. You want to connect the leads simultaneously.

5. Verify that the multimeter shows a net voltage of approximately 5V (see Figure 3.7).

FIGURE 3.7 You can use a multimeter and two PCB test points to quickly verify Raspberry Pi voltage.

You'll also need a *soldering iron*, which is a tool you use to permanently join electrical components and to extend circuits.

The very idea of soldering intimidates some people, but the kernel idea at play is really quite simple. You heat up the soldering iron to 700 degrees Fahrenheit or so and then melt solder into a junction between two other conductive components. When the solder dries, you have a permanent connection that allows electrical current to flow between the soldered components.

In a nutshell, *solder* is a fusible metal alloy that, as a conductor, can transmit electricity.

For Raspberry Pi projects, I recommend you get an adjustable 30W pen-style soldering iron for maximum flexibility. You also should purchase a spool of 60/40 lead rosin-core solder with a 0.031-inch diameter. The 60/40 means that the solder consists of 60 percent tin and 40 percent lead.

To make your soldering experience as user-friendly as possible, you might also want to look into the following relevant soldering accessories:

- **Solder sucker**: This vacuum tool makes short work of removing melted solder particles.

- **Solder wick**: This material, also called *desoldering braid*, is used in conjunction with the solder sucker to remove solder from your components.

- **Soldering stand**: We have only two hands, and if you are operating the iron with one hand and the solder wire with the other, then how the heck can you position your components to be soldered? Because using The Force isn't an option (probably), a soldering stand makes this easy.

Raspberry Pi Starter Kits

With so many hardware options available to Raspberry Pi enthusiasts, figuring out where to start can seem overwhelming to even experienced tech junkies.

For the befuddled (among whose number I once counted myself a member), several third-party vendors assemble so-called *Raspberry Pi starter kits* that include everything you need to start building Raspberry Pi-based projects. Heck, most of the kits even include a Model B board! This section lists a few such kits.

The *Adafruit Raspberry Pi Starter Pack* (http://is.gd/cf3ply) lists for $104.95. Besides the Model B, you also get the following components for your money:

- Adafruit clear acrylic Pi box
- 3' long Micro USB cable
- 5V 1A power adapter
- USB TTL console cable
- 4GB SD card
- Assembled Adafruit Pi Cobbler kit with GPIO cable
- USB microSD card reader
- Large, full-size breadboard
- Breadboarding wires
- 10' long Ethernet cable
- Embroidered Raspberry Pi badge
- 5 x 10K ohm resistors
- 5 x 560 ohm resistors
- 1 Red 10mm diffused LED
- 1 Green 10mm diffused LED
- 1 Blue 10mm diffused LED
- 3 tactile pushbuttons
- Light-sensitive resistor photocell
- 1 x 1uF capacitor

The *Maker Shed Raspberry Pi Starter Kit* (http://is.gd/YIWK7Z) costs $129.99 and also includes a Model B board. In addition, the pack includes the following:

- Adafruit Pi Cobbler Breakout Kit
- MAKE: Pi Enclosure
- 2-port USB wall charger (1A) and USB cable
- 4GB Class 4 SDHC flash memory card
- Deluxe, full-sized breadboard
- Mintronics Survival Pack with 60+ components, including voltage regulators, trimpots, LEDs, pushbuttons, battery snaps, capacitors, diodes, transistors, and resistors
- HDMI high-speed cable, 1.5 feet
- Deluxe breadboard jumper wires

The previous two Raspberry Pi starter kits are decidedly weighted toward electronics experimentation. There are also vendors who will sell you kits that include the Pi as well as core peripherals.

For instance, consider the *CanaKit Raspberry Pi Complete Starter Kit* ($89.95, http://is.gd/HnsiOJ), which includes the following parts:

- Raspberry Pi Model B
- Clear case
- Micro USB power supply
- Preloaded 4GB SD card
- HDMI cable

MCM Electronics (http://is.gd/6cAfs3) offers several different Raspberry Pi kits. Their *Raspberry Pi Enhanced Bundle* lists for $79.99 and includes the following:

- Raspberry Pi Model B
- 5V 1A Micro USB power supply
- Raspberry Pi case
- 4GB SD card preloaded with Raspbian (We discuss Raspbian in exhaustive detail in Chapter 4, "Installing and Configuring an Operating System.")
- 88-Key USB mini keyboard
- USB optical mouse

I mentioned in the previous chapter that you should consider purchasing a case for your Raspberry Pi. This is important for the following reasons at the least:

- Protection against electrostatic discharge (ESD).
- Protection against physical damage (remember that the Pi PCB is delicate).
- Some of the cases make the Pi look cool!

Although all cases I've seen fit both Model A and Model B (after all, both models use the same PCB), I suggest you purchase cases specific to each model. You'll find that Model B cases with Model A hardware create big gaps into which dust and other detritus can easily reach the delicate inner components.

Also as I said in the last chapter, be sure to select a case that gives easy access to the GPIO pins (and potentially the CSI camera interface) from outside the case. You want to provide yourself with as much flexibility as possible with your Pi.

Next Steps

If you've been reading this book sequentially thus far (and I sincerely hope you have), you should have a crystal-clear picture of what is necessary to get started with Raspberry Pi development.

You've heard the age-old adage, "You've got to learn to crawl before you can walk." Well, what this means for our purposes is that you need to learn how to write computer programs before you can link your Pi to an expansion board and start to perform meaningful work. Before we even get to writing code with Scratch and Python, you need to get comfortable with Linux. That's the subject we tackle in Chapter 4.

Installing and Configuring an Operating System

The operating system (also called the OS, pronounced *oh-ess*) is the most important software on your computer because it represents the interface between you, running software programs, and the machine's underlying hardware.

To put this discussion in a more meaningful context, let's say that you want to send an email message from your computer or mobile device. You compose and send your message by using a web browser or a dedicated email application. The computer's operating system translates your keystrokes and mouse movements into instructions that are understood by your application.

When you click Send, the OS then conducts the message through a number of subsystems, including translating your message data into an intermediate format that your computer's network interface card (NIC) device driver software can process.

The NIC driver and the networking protocol stack further segments the email message and ultimately transmits binary digital data in the form of 0s and 1s across the network communications media through the NIC hardware.

Figure 4.1 shows the conceptual relationship between the user, the application, the operating system, the device driver, and the device itself.

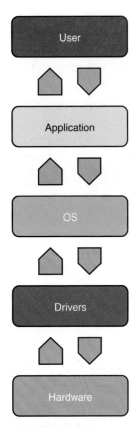

FIGURE 4.1 The operating system orchestrates data flow between the user, applications, device drivers, and the underlying computer hardware.

Let me briefly explain the layers shown in Figure 4.1:

- **OS**: The operating system orchestrates the communication among all the other layers.
- **User**: The computer operator provides input to and received output from the computer.
- **Application**: The program or application enables the user, operating system, and hardware to get work done.
- **Drivers**: Each hardware device needs platform-specific software that enables the OS to interact with it. In Linux, the OS kernel often integrates device drivers into its code.
- **Hardware**: Hardware can be either permanently part of the computer (like a motherboard) or an attached peripheral (like a mouse).

Like any other computer, the Raspberry Pi includes an operating system as well. Therefore, the OS skills that you'll pick up in this chapter are abilities that you will apply every single day in your work with the Pi. In later chapters we'll apply your Linux operating system skills as we undertake various Pi-related projects.

Common Operating Systems

Truly, operating systems are not at all voodoo magic, although their underlying structure can sometimes get complicated. The bottom line is that any end user of a computer system interacts with an operating system of one type or another.

For instance, a Mac user deals with the Apple OS X (pronounced *oh ess TEN*, not *oh ess EX*) operating system. A PC user generally uses a version of the Microsoft Windows operating system—for example, Windows 8 or Windows 7. Mobile device users might run one of the following operating systems, depending on the make and model of their mobile hardware:

- Apple iOS
- BlackBerry
- Google Android
- Windows Phone
- Windows RT

You might know already that the Raspberry Pi runs none of these. What's going on here? Well, as it happens, the RasPi uses a particular distribution of Linux.

Understanding Linux

Linux (*LIH-nix*) is an open source operating system originally developed by the Finnish computer scientist Linus Torvalds (*LEE-nus TUR-valds*) in 1991. Linux was built from the ground up as a free operating system that any interested party could tinker with, improve upon, and re-release under the GPL license (discussed back in Chapter 1, "What Is the Raspberry Pi?").

At some point you might have heard a nightmare story or two about how difficult it is to use. There was a time when the only people who would touch Linux were computer science nerds or grizzled government employees. Believe me, friends: Linux has gotten much more user-friendly over the past 10 years or so.

Thanks to Linux vendors like Red Hat and Canonical, Linux has become much more mainstream, often to the chagrin of the original Linux fanatics who prefer a less corporate, structured approach to OS development and distribution. In point of fact, Canonical's Ubuntu Linux, in its 12.10 version, bears a striking resemblance to Apple OS X (see Figure 4.2).

FIGURE 4.2   Apple OS X above, and Ubuntu Linux below.

The OS X and Ubuntu Linux user interfaces are quite similar indeed. As you will learn pretty soon if you haven't already, I'm a big believer in using (brief) unordered lists to teach concepts. Allow me to summarize what I see as the chief advantages of Linux:

- Generally more secure than proprietary OS software such as Windows and OS X because the community quickly squashes bugs and vulnerabilities.

- Gives the operator control of every aspect of OS operation, right down to the bare kernel level.

- The OS and most available software are free.

- You can do the vast majority of stuff in Linux that you are accustomed to doing in Windows or OS X.

In addition to these factors, it's also true that people who write malicious software tend to target the most popular operating systems simply because there are more available targets. Consequently, the relatively "niche" status of Linux in the consumer/enthusiast environment gives the platform a security advantage over mainstream OSs.

NOTE: BUT WHAT ABOUT OFFICE?

Long-time Linux users typically gravitate toward OpenOffice (http://is.gd/AxqDKr) or LibreOffice (http://is.gd/ORAFcy) as open source (and therefore free) alternatives to the proprietary Microsoft Office productivity suite.

As Bret Michaels sang in the 1980s, "Every rose has its thorn...." Here are what I see as the essential disadvantages of Linux:

- To access the raw power of Linux, you must learn how to use the command line, which involves a number of highly cryptic command-line tools.

- The graphical user interface in Linux is generally not as polished or intuitive as, say, Windows 7 or OS X.

- Configuring driver support for new hardware is sometimes problematic to Linux beginners due to the common requirement of manual driver installation and configuration.

- You can run Windows or OS X apps under Linux; however, doing so is not considered to be a beginner-level task. In general, the variety of software that is available to Linux is far less than what is available to, say, Windows or OS X.

Despite the challenges that running Linux has for us, I submit that Linux is truly the ideal operating system platform for the Raspberry Pi. Remember that the Pi is intended as a learning environment—what better way to discover the relationship between an OS and hardware than in an open-source situation where the underlying source code and hardware schematics are freely available to you?

Also, as we'll see momentarily, you have quite a bit of flexibility in terms of which Linux distribution you might prefer to run on our Pi.

Linux and Raspberry Pi

Remember when I said earlier that Linus Torvalds gave us Linux as a platform for community development? We call those Linux variations, those that come from the development community, *remixed* or *forked distributions*.

The Raspberry Pi Foundation put together an official Linux distribution that is optimized for Raspberry Pi; this distribution is called *Raspbian* (*RASS-pian*). The name Raspbian bears a bit of an explanation. Raspbian is a *portmanteau*, which is a mash up of two or more words derived from two separate technologies:

- **Raspberry Pi**: The $25/$35 computer upon which this book is based
- **Debian**: The Linux distribution used as a base for Raspbian (http://is.gd/lgF8Ft)

Personally, I'm overjoyed that the Foundation used Debian as a base Linux for the Pi. Number one, Debian includes one of the most powerful and flexible package managers in the industry (more on that in the next couple chapters). The Raspbian user interface is shown in Figure 4.3.

FIGURE 4.3 The Raspbian Linux distribution includes the LXDE graphical user interface (GUI).

Number two, Debian is one of the more user-friendly Linux distros in existence. As a matter of fact, Ubuntu Linux is also based on Debian.

But can we go ahead and install the "real" Debian or Ubuntu on the Pi? Unfortunately, no—at least not without some major kernel hacking. Remember that the Pi board uses an ARM CPU. Most desktop computers today, at least in the retail space, use the Intel processor. As of this writing in spring 2013, neither Debian nor Ubuntu Linux supports the ARM processing architecture.

The Kernel and Firmware

If the operating system constitutes the software "body" of a computer system, then the kernel represents the brain. Specifically, the *kernel* is the OS subcomponent that functions most intimately with installed hardware devices.

What's cool about Linux is that you can customize and recompile the kernel to suit different situations. For instance, the Raspberry Pi Foundation modified the Debian Linux kernel to accommodate the ARM processor and other components included on the Pi board.

The Linux kernel is called *firmware* because it is software that is semi-permanently written to the first partition of your Raspbian SD card. I say semi-permanently because the firmware data persists after you power down the computer. However, you can update the firmware to a more recent version if need be.

We can contrast data that is stored on the SD card with data that is stored in random access memory, or RAM. RAM-based data persists only as long as the Pi is powered up; unless you save RAM contents to the SD card, that data is permanently lost if the Pi is turned off or rebooted.

Raspberry Pi uses its own custom-built firmware that "blends" the proprietary Broadcom BCM2835 system on a chip (SoC) with the Raspbian operating system. In point of fact, the BCM2835 SoC actually has two sets of firmware flashed onto the SD card. The first is responsible for managing the hardware resources on the Pi, and the second is charged with controlling the behavior of the Pi's graphical processing unit (GPU).

NOTE: I'M FLOATING!

The original Raspbian code was not optimized in the kernel to process floating-point (decimal) numbers in hardware. This "soft float ABI (application binary interface)" situation, which involves emulating math co-processing in software, bothered experienced Linux users who wanted to use the Pi to perform more complex math. Fortunately, the current versions of Raspbian now contain a "hard float" ABI, which means that instructions for processing floating-point numbers are performed in hardware using the math co-processor chip. Needless to say, hard float is orders of magnitude faster than soft float.

Updating the Raspberry Pi kernel firmware is covered in the next chapter.

Raspberry Pi's Other Operating Systems

Keeping in spirit with the "do it yourself" philosophy of Linux, you can run a number of specially crafted Linux distros on the Pi. Raspbian is considered to be the reference operating system because it was built from the ground up for learning software and hardware programming with the Pi board. However, alternatives exist that are optimized for other uses. Let's take a brief look at a few of them:

Arch Linux ARM (http://is.gd/6EJlou): This distro is an ARM-specific branch of the Arch Linux OS that is aimed at experienced Linux users. Its structure is lightweight and is intended to provide the user with as much control as possible.

Fedora Remix (http://is.gd/Nj0Iys): This distro is an ARM port of the highly successful Fedora Linux OS. In particular, check out the Pidora distribution (http://is.gd/2TfKjx). Many Linux users swear by Fedora, so its ability to run on the Raspberry Pi pleases many enthusiasts.

Occidentalis (http://is.gd/t79m03): This distro, pronounced *ocks-ih-den-TAIL-is*, was developed by Adafruit and includes lots of OS "extras" to make hardware hacking easier. Adafruit is one of the best Raspberry Pi education sites out there; they sell extension hardware and provide detailed instructions on how to use it with your Pi.

OpenELEC (http://is.gd/KpaeqS): This distro, pronounced *open ee-LECK*, has a single aim—to run the Xbox Media Center (XBMC) as efficiently as possible. OpenELEC and XBMC are discussed in great detail in Chapter 12, "Raspberry Pi Media Center."

RaspBMC (http://is.gd/KyBKzy): This distro is like OpenELEC inasmuch as it is intended only to run Media Center software on the Pi.

RISC OS (http://is.gd/6EJlou): This distro, pronounced *risk oh ess*, was developed by Acorn, who you'll remember is the manufacturer of the BBC Micro microcomputer, the Raspberry Pi's inspiration.

Of these alternative Linux distros for the Raspberry Pi, I personally like Occidentalis the best because the environment is optimized for use with the Adafruit Learning System (http://is.gd/efFtD7). Be sure to visit and bookmark the Adafruit website; they offer almost every conceivable Raspberry Pi hardware add-on. Figure 4.4 shows the Occidentalis user interface.

Please note that the Raspberry Pi-compatible operating systems suggested here represent only part of what's available. Check out the RPi Distributions page at the Embedded Linux Wiki (http://is.gd/3yHQZ2) for a more complete rundown.

FIGURE 4.4   Adafruit's Occidentalis Linux distribution, which includes plenty of hardware hacking tools.

Installing Raspbian on Your Raspberry Pi

My first computer, the Tandy TRS-80 Model III, and the Commodore 64 had no persistent, onboard storage. That is to say, anything you wanted to save permanently, such as a BASIC program that took 12 hours to type in, had to be saved to some external media to survive a system shutdown or restart.

It was a great day in the early 1990s when Intel-based personal computers began shipping with fixed internal hard drives! Suddenly you could boot an operating system, load programs, and save data to your heart's content, free of the worry of losing all of your work when you turned off the computer!

Believe it or not, the Raspberry Pi board also does not contain an internal disk drive to boot the OS and save user data. Instead, the Foundation included an SD card slot.

If you read Chapter 3, "A Tour of Raspberry Pi Peripheral Devices," you have all the information you need to purchase your SD card. I formally suggest you stay with the Standard SD card form factor (see Figure 4.5), but you can make use of the Mini or Micro cards with the appropriate adapters.

Building a Raspbian SD card is not as easy as copying a bunch of files from a folder on your PC or Mac to the SD card. Instead, you download a binary OS image file and flash that image onto your SD card in one pass.

FIGURE 4.5   You can learn much about an SD card by studying its sticker label. Here we can see the brand, capacity, and speed rating in a single glance.

NOTE: WHAT TYPE OF SD CARD SHOULD I USE?

I suggest you purchase a name-brand SD card of at least 4GB capacity. Moreover, I recommend that the SD card speed be rated at Class 4 or higher. You can visit the Embedded Linux website (http://is.gd/Ym6onO) to view a comprehensive compatibility index of Raspberry Pi-compatible SD cards.

You'll need the following ingredients to create a Raspbian OS image SD card:

- PC or Mac computer
- SD card reader (some computers include these in the case)
- SD image burning tool or command-line equivalent
- Official Raspbian OS image (http://is.gd/6EJlou)

The following sections take you through the process of setting up an SD card with Raspbian.

TASK: CREATING A RASPBIAN SD CARD IN WINDOWS

Win32 Disk Imager, which you can get from http://is.gd/UkTdSW, is the recommended tool for building a Raspbian OS image SD card in Microsoft Windows. To begin, you'll need to download both it and your Raspbian distro. When those are in hand, follow this procedure:

1. Plug your SD card into your Windows computer and make a note of which drive letter Windows assigns to the device. It does not matter if there is any data already stored on the card—it will be overwritten (be careful!).

2. Use Windows Explorer or a ZIP file management utility (I like the free 7Zip from http://is.gd/oOJvG8) to unpack both the Raspbian OS image and Win32 Disk Imager.

3. Fire up Win32 Disk Imager. You'll note that all you have to do is unzip the package and run the executable program file—there is no installation.

4. Click the Browse button (marked 1 in Figure 4.6) and locate the Raspbian .img file that you downloaded from the Raspberry Pi Foundation website.

5. Open the drop-down list (marked 2 in Figure 4.6) and specify the drive letter that is associated with your mounted SD card. Again, be careful to ensure that you have the correct drive letter selected.

6. Click Write (marked 3 in Figure 4.6) to flash the card.

7. Close Win32 Disk Imager and eject your SD card. You're finished!

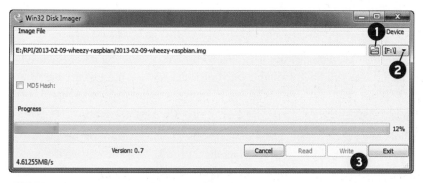

FIGURE 4.6   The open source Win32 Disk Imager is perhaps the best SD card flashing utility for the Windows OS.

TASK: CREATING A RASPBIAN SD CARD IN OS X

For OS X computers, we can use either the dd Terminal shell program or a graphical utility. In my experience, GUI SD flashing tools for OS X work sporadically if at all. Therefore, we shall use the tried-and-true command-line technique.

GUI-BASED INSTALL TOOLS

For those who are interested (and brave), here are a couple GUI tools that theoretically can be used to flash Raspberry Pi OS images to SD cards:

- RPi-sd card buider (http://is.gd/AJTQfM)

- Pi Filler (http://is.gd/8WplZ7)

1. Download your Raspbian OS image to your Mac, and double-click the ZIP file to extract the underlying .img file.

2. Plug your SD card into your Mac computer and start Terminal. The easiest way to do this is to open Spotlight, type **terminal**, and press Enter (see Figure 4.7).

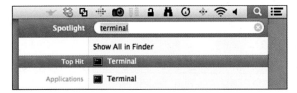

FIGURE 4.7 Spotlight provides you with an ultra-fast way to locate and run programs on your Mac.

3. From the Terminal prompt, type **diskutil list**. This command shows you all of the fixed and removable disk drives you have available on your Mac. Make a note of the device path for your SD card. For instance, you can see in Figure 4.8 that my brand-new 4.0 GB SD card is addressed as disk2.

If your SD card has been used before, there may already be one or more partitions defined on the card; they should be addressed as disk2s1, disk2s2, and so forth. It's important to note that we want to flash the entire contents of the card and not an individual partition.

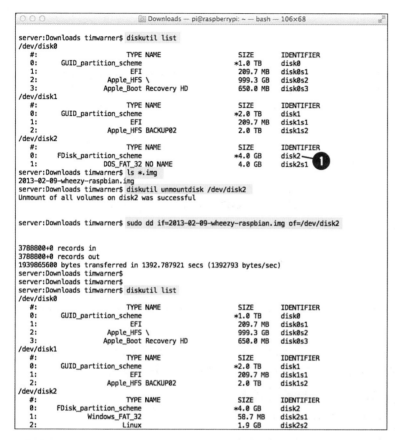

```
server:Downloads timwarner$ diskutil list
/dev/disk0
   #:                       TYPE NAME                    SIZE       IDENTIFIER
   0:      GUID_partition_scheme                        *1.0 TB     disk0
   1:                        EFI                         209.7 MB   disk0s1
   2:                 Apple_HFS \                        999.3 GB   disk0s2
   3:                 Apple_Boot Recovery HD             650.0 MB   disk0s3
/dev/disk1
   #:                       TYPE NAME                    SIZE       IDENTIFIER
   0:      GUID_partition_scheme                        *2.0 TB     disk1
   1:                        EFI                         209.7 MB   disk1s1
   2:                 Apple_HFS BACKUP02                 2.0 TB     disk1s2
/dev/disk2
   #:                       TYPE NAME                    SIZE       IDENTIFIER
   0:      FDisk_partition_scheme                       *4.0 GB     disk2
   1:                 DOS_FAT_32 NO NAME                 4.0 GB     disk2s1
server:Downloads timwarner$ ls *.img
2013-02-09-wheezy-raspbian.img
server:Downloads timwarner$ diskutil unmountdisk /dev/disk2
Unmount of all volumes on disk2 was successful

server:Downloads timwarner$ sudo dd if=2013-02-09-wheezy-raspbian.img of=/dev/disk2


3788800+0 records in
3788800+0 records out
1939865600 bytes transferred in 1392.787921 secs (1392793 bytes/sec)
server:Downloads timwarner$
server:Downloads timwarner$
server:Downloads timwarner$ diskutil list
/dev/disk0
   #:                       TYPE NAME                    SIZE       IDENTIFIER
   0:      GUID_partition_scheme                        *1.0 TB     disk0
   1:                        EFI                         209.7 MB   disk0s1
   2:                 Apple_HFS \                        999.3 GB   disk0s2
   3:                 Apple_Boot Recovery HD             650.0 MB   disk0s3
/dev/disk1
   #:                       TYPE NAME                    SIZE       IDENTIFIER
   0:      GUID_partition_scheme                        *2.0 TB     disk1
   1:                        EFI                         209.7 MB   disk1s1
   2:                 Apple_HFS BACKUP02                 2.0 TB     disk1s2
/dev/disk2
   #:                       TYPE NAME                    SIZE       IDENTIFIER
   0:      FDisk_partition_scheme                       *4.0 GB     disk2
   1:                 Windows_FAT_32                     58.7 MB    disk2s1
   2:                        Linux                       1.9 GB     disk2s2
```

FIGURE 4.8 Flashing a Raspbian OS image to an SD card in OS X. The relevant commands are highlighted in yellow.

4. Take the SD card offline by issuing the command **diskutil unmountdisk /dev/diskN**, where **N** is your SD card's device ID. As shown by 1 in Figure 4.8, my SD card is called disk2.

5. Navigate to the folder that contains your Raspbian OS image. Because I stored my .img file in my Downloads folder, I used the command cd Downloads to switch from my home directory to the Downloads directory. You then can specify the input file path of the dd command by using only the file name instead of with a full path.

6. Use the dd command to flash the Raspbian OS image to the SD card. Here is the statement:

```
sudo dd if=image_file_name.img of=/dev/diskXsY bs=2M
```

And here's a breakdown of each command in the statement:

- **sudo**: This command instructs the computer that you want to issue this command with administrator privileges. Your OS X user account has to be an administrator to complete this task.

- **dd**: This command, which officially stands for "Data Description" and unofficially stands for "Data Destroyer," is used to flash binary images to removable media.

- **if**: This command specifies the path to the input file. In this case, it is the Raspbian OS image.

- **of**: This command specifies the path to the output, which in this case is the SD card's target partition.

- **bs**: This command stands for block size, and larger values write more data chunks to the SD card in less time, but you are more likely to have errors. Best practice states that two megabytes (2MB) is a good compromise between speed and accuracy.

After the image flash completes, you'll see a confirmation message and get your command prompt back. You're finished!

TASK: CREATING A RASPBIAN SD CARD IN LINUX

Remember that Apple OS X has a core in Unix and that Unix formed the basis for all Linux distributions. Consequently, we can make use of the trusty dd command to flash our Raspbian image to our SD card in Linux. Here are the steps (*note*: I use Ubuntu Linux 12.10 as my reference desktop Linux distribution throughout this book):

1. Download the latest Raspbian OS image from the Raspberry Pi Foundation website. You can double-click the ZIP file to extract the .img file inside. Make a note of the location of the file.

2. Open up Terminal; in Ubuntu this is accomplished by clicking the Dash icon, typing **terminal**, and pressing Enter (see Figure 4.9).

   As you can probably tell, the Unity user interface in Ubuntu 12.10 takes some getting used to. When you click the Dash icon on the quick launch bar, a prompt window appears in which you can simply start typing for the program, document, or setting you want to find.

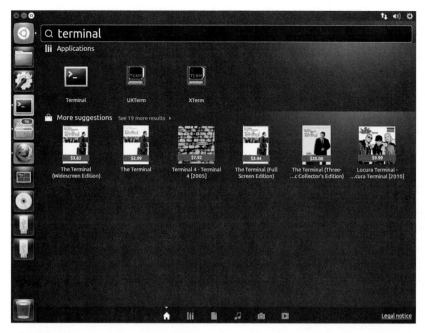

FIGURE 4.9   The Dash in Ubuntu 12.10 functions identically to Spotlight in OS X.

Almost immediately you'll see local and Internet-based results for your search query. In Figure 4.9, you can see that typing **terminal** brings back the built-in Terminal application as the first result.

As with the OS X flashing procedure earlier, the full input and output for the Linux process in is shown in Figure 4.10. Again, the commands are highlighted for your easier reference.

3. Plug in your SD card and wait a couple of minutes for Linux to detect your card. Next, run the command **fdisk -l** from a Terminal prompt. This step and the rest of the procedure is highlighted in Figure 4.10.

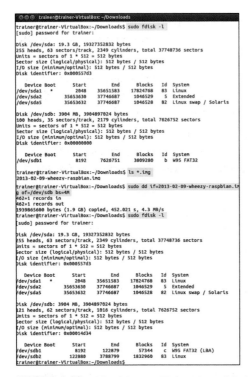

FIGURE 4.10 Flashing a Raspbian OS image to an SD card in Ubuntu Linux.

In the screenshot note the reference to the device path /dev/sdb1. In Linux, hard drives and removable media are mounted as file system paths.

CAUTION: BE CAREFUL!

I can't stress this enough: The dd command is extremely powerful, and if you don't watch which drive path you point it at, you could easily erase your boot drive and your valuable data!

4. Run the following statement:

```
sudo dd if=image_file_name.img of=/dev/disksdX bs=2M
```

Here is a breakdown of the syntax:

- **sudo**: Run the dd command under administrative privilege.
- **dd**: This command actually performs the OS flash.

- **if**: Here you specify the full or relative path to the OS image. In Figure 4.10, you see that the **cd** command is used to switch to the directory that contains the Wheezy OS image file.

- **of**: This command specifies the path to the output. This is important: Note in Figure 4.10 that /dev/sdb is specified as the target, not /dev/sdb1. You don't want to flash a specific partition on the disk—you want to flash the image onto the entire SD card.

- **bs**: This stands for block size, and larger values write more data "chunks" to the SD card in less time, but you are more likely to have errors. Two megabytes (2MB) is a good compromise between speed and accuracy. And yes, the syntax specifies M instead of MB for the block size value.

5. Run **fdisk -l** again and look at your /dev/sdb entries. You'll see that Raspbian creates two partitions on the SD card. One partition is a File Allocation Table (FAT) Windows-compatible partition. The other is much bigger and is a Linux partition.

An All-in-One Solution

Because the process of installing an operating system to the Raspberry Pi can be so tedious and scary for computing newcomers, some developers have taken it upon themselves to build tools to lessen this learning curve.

Take the New Out Of Box Software (NOOBS) project (http://is.gd/0n2yZv), for example. This is a tiny boot loader that makes installing an OS on your Raspberry Pi a breeze.

As you can see in Figure 4.11, the NOOBS user interface presents a simple menu from which a user can install any of the following Pi-tailored operating systems:

- Arch Linux
- OpenELEC
- Pidora
- RaspBMC
- Raspbian
- RiscOS

FIGURE 4.11 The NOOBS utility makes it much easier to load an OS on your Raspberry Pi.

After you've used NOOBS to install an OS on your Pi, the boot loader remains resident on your SD card such that you can reinvoke it at any time by holding down the Shift key during bootup.

TASK: INSTALLING AN OS ON YOUR RASPBERRY PI USING NOOBS

The procedure of setting up a NOOBS SD card differs a bit from the process we've used thus far in flashing an OS to the card.

1. Download the SD Card Association's formatting tool, SD Formatter (http://is.gd/IFMlmc), install the software, and use it to format your SD card. Remember to use an SD card of at least 4GB capacity.

   The SD Formatter utility formats the SD card such that we can interact with the drive directly in Windows or OS X.

2. Download NOOBS (http://is.gd/6EJlou) and extract the contents to the root level of your newly formatted SD card. Note that we aren't flashing a binary image to SD like we've done previously in this chapter.

   To repeat: We are manually copying the NOOBS file contents directly to the SD card.

3. Insert your newly prepared SD card into the Pi and boot it up.

4. When the Pi Recovery window appears, select your desired operating system and click Install OS.

When the OS installation completes, the Pi reboots and automatically loads the chosen OS. You're done!

If nothing else, NOOBS provides a quick, convenient, and easy way to test out a number of different OSs on the Pi to help you find your "comfort zone."

Testing Your New Raspbian Image

The best way to verify that your newly created Raspbian image works correctly is to insert it (carefully) into the SD slot on your Pi's circuit board and plug the Raspberry Pi into your power supply. You'll also want to plug in a keyboard, mouse, and monitor/TV, as discussed in Chapter 2, "Hardware Components Quick Start."

If all goes well, you'll see the Raspi-Config interface as shown in Figure 4.12.

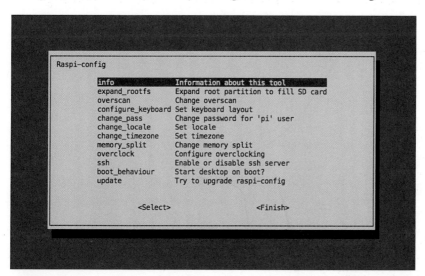

FIGURE 4.12   First-time Raspberry Pi setup is accomplished by using the Raspi-Config script.

If for some reason the Raspberry Pi doesn't boot directly into Raspi-Config, then simply issue the command **sudo raspi-config** from the Terminal prompt.

You learn the details of Raspberry Pi first-time configuration next in Chapter 5, "Debian Linux Fundamentals—Terminal."

Debian Linux Fundamentals—Terminal

In general, people tend to be afraid of the Linux operating system. In my experience, the following represent the most common complaints:

- Linux is at least partially command line, and that is intimidating.
- Linux uses all these wacky commands with strange syntax.
- Even the Linux graphical environments are nothing like what we are used to with Windows or OS X.

Certainly there is some truth to these concerns. Until a few years ago, Linux was in fact a seemingly impenetrable operating system to all but the most propeller-headed of computer geeks. However, the twenty-first century brought with it Linux distributions like Ubuntu and Raspbian, which are aimed at the ordinary computer user.

Sure, you need to learn some funky command-line syntax in order to get around Debian Linux on your Raspberry Pi. Nevertheless, I believe that you'll be much better off for it. You might even find yourself turning to the command line in Windows and OS X when you see how much quicker you can get work done via the keyboard.

Baby Steps

Before you can begin using the Debian Linux command line, you need to arrive at a command prompt on your Raspberry Pi. If you connect directly to your Pi (that is to say, by using a monitor, keyboard, and mouse), you should land at a command prompt by default.

You'll be required to log into your Raspberry Pi: The default username is pi, and the default password is raspberry. Press Enter after inputting each part of your logon.

NOTE: ON CASE SENSITIVITY

Raspbian, like all Linux distributions, is case-sensitive. You might already be accustomed to the fact that passwords are case-sensitive, but it may take some getting used to things like having two files named File.txt and file.txt co-existing in the same directory!

If you connect to your Pi remotely using SSH or VNC (procedures covered fully in Chapter 7, "Networking Raspberry Pi"), you must establish that remote connection before accessing the Debian command prompt.

Finally, if you booted your Pi into the LXDE graphical environment, you can open a command prompt (formally called the Terminal) by double-clicking the LXTerminal icon on the Desktop. This is shown in Figure 5.1.

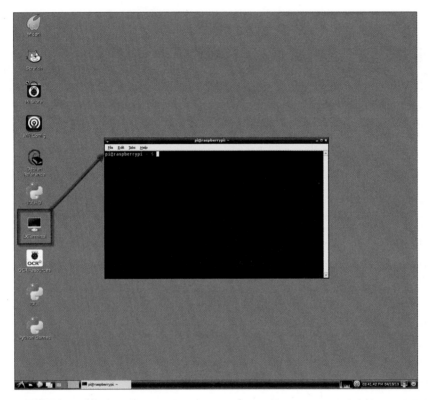

FIGURE 5.1 You can start a Terminal session from within the LXDE desktop environment.

Regardless of you how you get to the Raspbian command prompt, the end result is the same; namely, you see an input area that looks like the following:

```
pi@raspberrypi ~ $
```

The command prompt itself actually yields highly valuable system information. Let me break it down for you element by element:

- **pi**: This is the name of the default Raspbian user account.
- **@**: This denotes the link between the currently logged on user and the computer name.
- **raspberrypi**: This is the default hostname (computer name) for your Raspberry Pi.

- ~: This denotes your current location in the Debian filesystem. The tilde (~) represents a shortcut representation of the user's home directory, which has the full path /users/raspberrypi in Raspbian.

- $: The dollar sign denotes a nonadministrative user account. By contrast, when you switch your user identity to the root (superuser) account, the prompt changes to the octothorpe (#) character.

NOTE: ON NAMING CONVENTIONS

You'll notice that in this chapter I use the terms Debian and Raspbian interchangeably. Don't be confused! Everything you learn in this chapter works 100% in any distribution of Debian Linux. Therefore, it doesn't matter whether you are testing these procedures in Raspbian, Debian, Ubuntu, or another Debian-derived Linux distribution.

Linux provides the operator full control over the environment. To that end, you can customize the format of the command prompt in myriad ways. If you'd like more information on this, please read this post from the Lindesk blog: http://is.gd/xFxt2f.

The Linux command-line interface (CLI, pronounced *see-ehl-eye*) is what is called a *command shell*, or simply *shell*. Specifically, Debian employs the Bourne Again Shell (Bash) by default. As with anything and everything else in Linux, you can swap out the Bash shell for another more to your liking. For our purposes in this book, I stick to Bash.

Essential Terminal Commands

When you are logged into a Raspbian Terminal, where can you go from there? Before we go any further, I want to provide you with the core Terminal commands that you should know in Raspbian (Debian) Linux.

Having these commands in your tool belt immunizes you against getting "stuck" with your Raspberry Pi. A good example of this is the common newcomer situation of not knowing how to go from the Bash shell to the LXDE graphical environment (and vice versa).

startx

Many Raspberry Pi enthusiasts don't want to mess with the command shell any more than they absolutely have to, instead preferring to work in the GUI environment. To jump from the command line to the GUI, simply type **startx** and press Enter.

When you are in the LXDE GUI environment, you can open the LXTerminal application as previously mentioned to spawn a new command-line Bash session. If you want to unload the GUI and return to a fully character-based interface, simply click the Logout button in the bottom-right corner of the LXDE Desktop. Next, in the message box that appears, click Logout (see Figure 5.2).

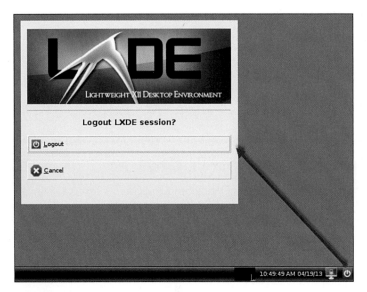

FIGURE 5.2   You can easily return to a full-screen Bash prompt by using the Logout command in LXDE.

pwd

The present working directory (pwd) command answers the question "Exactly where am I in the Raspbian file system?"

I suppose it would be helpful to provide a brief description of the Linux file system. Your Raspberry Pi hard drive is represented as a hierarchical tree of directories (folders) and files. The top level of the file system is denoted with a slash (/) character.

Beneath the root are a bunch of standard directories, some of which are more important to the end user than others. Here's a quick breakdown:

- **bin**: Essential programs
- **boot**: Boot loader programs (that is, files necessary to boot the Pi)
- **dev**: Hardware device drivers and information
- **etc** (pronounced *etsy* or *ee-tee-see*): System-wide configuration files
- **home**: Users' home directories (personal settings, data files, and so forth)
- **lib** (pronounced *lihb*): Code libraries that are used by the system programs in /bin and /sbin
- **lost+found**: File fragments picked up by file system repair tools like fsck
- **media**: Mount points for removable media (DVDs, flash drives, USB portable drives)
- **mnt**: Temporarily mounted file systems

- **opt**: Optional application software packages

- **proc** (pronounced *prock*): Dynamically changing system status information

- **root**: Home directory for the root user account

- **run**: Supplemental runtime data stored by installed applications

- **sbin** (pronounced *ess-bin*): Executable program files that are reserved for administrative use

- **selinux**: Working directory for Security Enhanced Linux, a security enhancement toolset

- **srv**: Temporary storage for some services such as File Transfer Protocol (FTP)

- **sys**: Stores Linux operating system files

- **tmp**: Temporary files that are purged during every system reboot

- **usr** (pronounced *user* or *you-ess-arr*): Read-only user data; multiuser utilities and applications

- **var**: (pronounced *vahr*). Variable data whose values change over time (logs, spool files, temporary files, and so forth)

NOTE: WATCH YOUR PRONUNCIATION!

The main reasons I offer you pronunciations for many of these Linux-oriented terms is because (a) The acronyms are often difficult to pronounce at any rate; (b) Some Linux power users get awfully persnickety about correct pronunciations; and (c) I want to equip you with all the tools, physical, logical, and verbal, to become a proficient Linux user.

ls

After you've figured out where you "live" in the Raspbian file system, you probably want to see the contents of that present working directory. That's what the **ls** command does; it runs a directory listing.

In my experience, **ls** will be one of your most frequently used commands. After all, you need some mechanism of visualizing directory contents from a command prompt.

Terminal commands often employ switches or parameters to customize how the command works. For instance, try running the following:

```
ls -la
```

This changes the output quite a bit, doesn't it? Take a look at Figure 5.2 to see the command output on my Pi. The l switch gives you a columnar (or long; hence the l) listing. The a switch shows all files, even hidden and system files.

You can even run directory listings for other directories on your Linux system (or, for that matter, attached storage devices). You simply append the relevant directory path to the ls command. For instance, check out the following example, the output for which is shown in Figure 5.3.

```
ls -la /usr/bin
```

The previous command shows you the contents of the /usr/bin directory no matter what your present working directory might be.

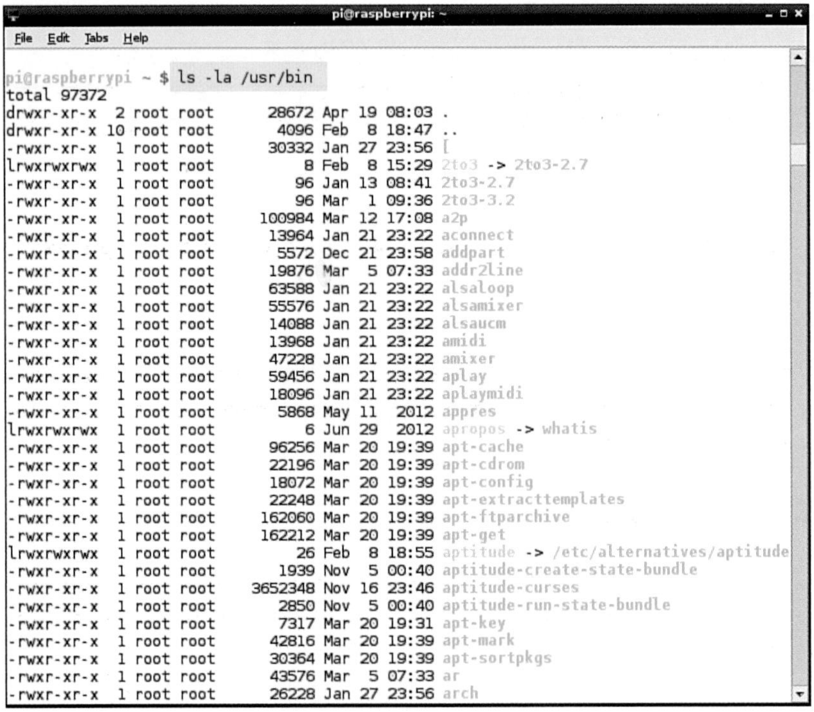

FIGURE 5.3 The ls command will be one of your most frequently-run Linux Terminal commands.

cd

The cd, or *change directory*, command is used to navigate the Linux file system from a command prompt.

If you know your destination ahead of time, simply supply the full path:

```
cd /Users/raspberrypi/Downloads
```

You can also use relative paths, which are partial file/directory paths that are built on the present working directory location.

For instance, type **cd** with no additional arguments to quickly return to your home directory. Then try the following:

```
cd Downloads
```

Be sure not to include a leading slash (/) before Downloads and remember that Downloads is case-sensitive.

TIP: TAB COMPLETION

As you type commands, paths, and file/program names in Raspbian, try hitting TAB to see if the Linux autocompletion feature is attractive to you. Believe me, tab completion comes in handy when you need to type in super-long and super-cryptic file names!

From the ~/Downloads folder, type the following to move up one level from your present working directory:

```
cd ..
```

Make sure to put a space between the cd and the two periods. Also, the tilde (~) character is a shortcut representation of the currently logged-on user's home directory path. Thus, for the Pi user the following paths can be used:

```
cd ~/Downloads
cd /Users/raspberrypi/Downloads
```

Don't be ashamed to run the pwd command frequently as you cd your way throughout the Raspbian (Debian) Linux file system. It's easy to get lost even if you do have an informative command prompt configured!

sudo

The sudo (pronounced *sue-doo*) command is one of the most important commands for you to know, not only for Raspbian, but for any Linux or Unix operating system (and that includes Apple OS X).

It is widely (and correctly) considered to be a security problem to actually run Linux under the context of the root (superuser) account; therefore, while remaining under standard account privileges, you can use sudo to temporarily elevate them and run administrative-level commands while remaining under standard account privilege otherwise.

To use sudo, you simply prepend the word before the command you want to run. The following example opens the hosts system configuration file in the nano editor as root:

```
sudo nano /etc/hosts
```

Historically, sudo as a command name is a portmanteau (word mashup) of two other words:

- **su**: This is a Linux command that means *substitute user* and is used to change the current user account associated with the current Terminal session.

- **do**: This is nothing more than a reference to the verb do, which means to perform a particular action.

passwd

You use the **passwd** command to change the password for a user account. To change the default pi account password, issue the following Terminal command:

```
sudo passwd pi
```

You'll be asked to (1) authenticate with the current password; (2) define a new password; and (3) confirm said password.

If you have root privileges on the computer (which the pi account does by default), you can change the password for any user on the system as well.

NOTE: BECOMING ROOT

You can customize which Raspbian user accounts can employ the sudo command by making or editing entries in the /etc/sudoers system configuration file.

NOTE: TAKING THE NEXT STEP(S) WITH LINUX

Although I do my best in this book to give Linux newcomers what they need to become at least moderately proficient users, there is only so much space to work with. Therefore, I suggest you pick up a good book on Linux end user fundamentals (such as *A Practical Guide to Linux Commands, Editors, and Shell Programming, 3rd Edition,* by my Pearson colleague Mark Sobell: http://is.gd/NWMLHz).

nano

Linux distributions generally include several different text editors, although some long-time Linux users will balk at my suggesting nano (pronounced *NAH-noh*) instead of vi (pronounced *vee-eye* or *vie*).

NOTE: LINUX HUMOR

Linux program names often have colorful and/or ironic histories. For our purposes, nano is a recursive acronym for Nano's ANOther Editor. Historically, nano is a more user-friendly successor to an ancient Linux email client application named pico (pronounced *pee-koh*).

To open an existing file (such as /etc/hosts), run the following command:

```
sudo nano /etc/hosts
```

If you want read/write access to system configuration files like hosts, you should always prepend your nano command with sudo.

You can create a new, blank text file by running nano with the name of your new, as-yet-uncreated file. For instance, the following command creates a new file named test.txt in the present working directory and opens the document for editing in nano:

```
sudo nano test.txt
```

One thing I like about nano is that the primary interface commands appear in the footer of the user interface (see Figure 5.4).

FIGURE 5.4 The nano text editor's user interface includes onscreen navigation help.

In a nutshell, you use the arrow keys to navigate and the Control key to issue shortcuts. The most common of these shortcuts are Control+O to write out (save) your file, and Control+X to exit nano.

man

Any self-respecting Linux distribution includes a local library of manual (man) pages that describe the full purpose and syntax of Linux commands. Raspbian is no different!

To look up syntax for a particular Linux command (let's start with ls as an example), try this:

```
man ls
```

The man pages open by default in the page viewer less (run man less to learn more about this program!). You can see the screen output in Figure 5.5.

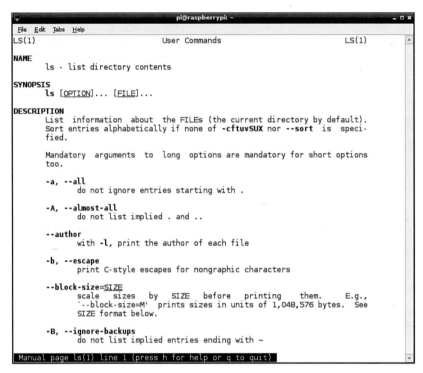

FIGURE 5.5 Raspbian man pages open in the less page viewer by default.

You can actually run **man man** to learn more about the man command itself.

To navigate a man page in the less viewer, use the spacebar to move one screen at a time and press Enter to scroll one line at a time. To exit the viewer, simply type **q**.

shutdown

In Debian, the shutdown command is my preferred way for not only shutting down the Pi, but also for performing reboots.

To initiate an immediate shutdown on your Pi, simply issue the following command:

```
sudo shutdown -h now
```

The -h parameter instructs Raspbian to halt the system as opposed to simply putting the system in a standby state. To *halt* Linux means to power off the machine entirely.

Somewhat ironically, you can also use the shutdown command to restart the Raspberry Pi. To do this, include the -r parameter as shown here:

```
sudo shutdown -r now
```

You will observe my use of sudo for any issuance of the shutdown command. In Linux, shutting down or restarting the system is a privilege reserved only for those with superuser (root) abilities.

The now parameter can be substituted with a time value if, for whatever reason, you want to delay a halt or a reboot. Consider the following example, which employs a 10-minute delay as well as a pop-up message to all connected users. Please note that the following code should be typed on a single line; this isn't two separate statements.

```
sudo shutdown -h +10 "Server is going down for maintenance. Please save your
work and logoff. Thank you."
```

You might have wondered, "Hey, Tim—where are the other file-management commands?" Those of you with some previous Linux experience probably have used one or more of the following Terminal commands:

- **cp**: Copy file
- **mkdir**: Make a directory
- **mv**: Move or rename a file
- **rm**: Remove a file
- **rmdir**: Remove a directory

In my experience, most file management tasks are more easily accomplished from the GUI shell as opposed to from the command line. That said, both methods are covered in the next chapter.

Updating Your Software

In Chapter 4, "Installing and Configuring an Operating System," you learned how to flash your SD card with the Raspbian operating system. More specifically, you observed post-flash that the disk contained not one but two partitions. The first partition contains the Raspberry Pi firmware, and the second contains the Raspbian operating system proper. You can see this disk layout in Figure 5.6.

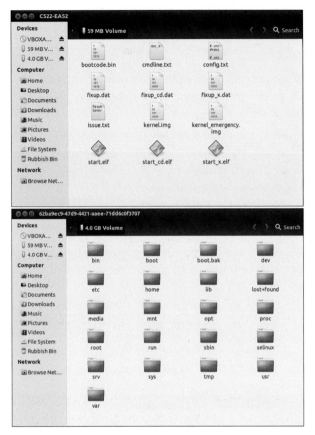

FIGURE 5.6 A Raspbian SD card contains two partitions: one for the Linux firmware and the other for the OS itself. The firmware files are shown in the top window, and the OS files are shown in the bottom window.

For security and stability reasons, it is important that you periodically run an update check not only for the operating system, but also for the firmware itself.

Updating Raspbian

Two Terminal commands need to be issued to update this Raspbian distribution. Here's the first command:

```
sudo apt-get update
```

Debian Linux distributions use the Advanced Packaging Tool, or APT, to locate, download, install, and remove application and OS software. The statement apt-get update fetches the latest updates to the Pi's configured software repository list.

In Linux parlance, a software repository is an online location where you can search for and install software for your computer. It's important to note that apt-get update does not actually update anything; it simply makes sure that your system has the latest and greatest info regarding software versions in the software repositories.

To perform an update operation, issue the following Terminal command:

```
sudo apt-get upgrade
```

To make this process more convenient, you can use the double ampersand (&&) concatenation operator to chain the two commands into one:

```
sudo apt-get update && sudo apt-get upgrade
```

We revisit the concepts of software repositories and the APT system in Chapter 6, "Debian Linux Fundamentals—Graphical User Interface."

Updating the Pi Firmware

As it happens, a member of the Raspberry Pi community named Hexxeh created a tool called rpi-update (http://is.gd/Rrh2bS) to automate the firmware update process. Here's the procedure:

1. Get to a Terminal prompt on your Raspberry Pi.

2. Run the following commands:

```
sudo dpkg-reconfigure tzdata
sudo apt-get install git-core
```

The dpkg statement checks the system's currently configured time zone. Git is an open source software version control application that is very popular among Linux developers.

3. Run the following command:

```
sudo wget http://goo.gl/1BOfJ -O /usr/bin/rpi-update && sudo chmod +x /usr/bin/
rpi-update
```

Wow—that was a long, honkin' command, wasn't it? Let's break down what it does:

- **wget** (pronounced *double yew get*): This is a Linux tool you use to retrieve web server content from Terminal (as opposed to from a web browser).
- **http://goo.gl/1BOfj**: This is simply a shortened Uniform Resource Locator (URL) to where the rpi-update program lives on github.com.
- **&&**: The double ampersand is a Linux concatenation operator that allows you to run one statement immediately after the preceding one completes.
- **chmod** (pronounced *see-aych-mod*): This is a Linux command you use to edit the permissions and attributes on files and directories. In this case, you are allowing rpi-update to run as an executable program file.

4. To perform a firmware update check/install, simply run the following command:

```
sudo rpi-update
```

If you do get a firmware update, you'll need to reboot your Pi. Remember that you can do that from the Terminal prompt by using the following statement:

```
shutdown -r now
```

Revisiting Raspi-Config

By far the easiest way to perform initial setup of your Raspberry Pi is to use the built-in Raspi-Config utility. You can start Raspi-Config from the full-screen Bash environment or from the LXTerminal simply by typing the following:

```
sudo raspi-config
```

Doing so presents you with the Raspi-Config text-based interface you saw in Chapter 4. Raspi-Config is actually a user-friendly front end to the config.txt configuration file located in the /etc directory. Figure 5.7 shows you what the Config.txt file looks like.

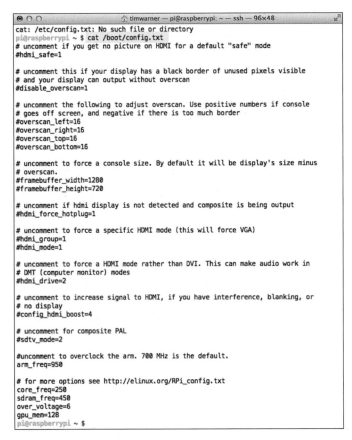

FIGURE 5.7 The Raspi-Config utility makes "under the hood" changes to the Config.txt file.

You can see the Raspi-config interface proper in Figure 5.8.

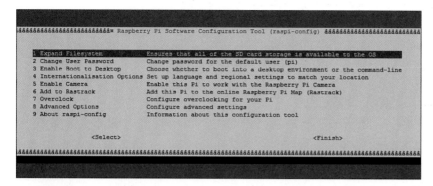

FIGURE 5.8   The Raspberry Pi Software Configuration Tool (Raspi-Config).

You can navigate through Raspi-Config utility by using the keyboard and can use your Up and Down arrow keys or the Tab key to move through the options; press Enter to make a selection.

To leave the main menu and leave Raspi-Config, use the Right Arrow or Tab until you've highlighted Finish, and then press Enter.

Let's close this chapter by walking through each Raspi-config option in greater detail.

Expand Filesystem

You should run this option as soon as possible after flashing your Raspbian SD card. Doing so makes the full space of your SD card available to Raspbian. If you don't expand the root file system, your Raspbian OS will be limited to a 2GB partition. This result is, of course, quite undesirable if you are using a 32GB SD card!

Change User Password

Everybody with any degree of familiarity with the Raspberry Pi knows that the default username is pi and the default password is raspberry. Thus, if you have any notion of storing confidential data on your Pi, you are best advised to change the password for the pi account immediately.

Remember you can also change the current user's password at any time by issuing the passwd Terminal command.

You learn how to create additional user accounts in the next chapter.

Enable Boot to Desktop

This option can be used to instruct the Pi to book directly into LXDE instead of stopping at the Terminal prompt. Of course, you need to perform some additional work if you habitually connect to your Pi remotely like I do. At any rate, Chapter 7 gives you the full skinny on Raspberry Pi connectivity options.

Internationalisation Options

This option opens a submenu that enables you to localize your Pi to its corresponding geographic location. Inside this submenu are three options:

- **Change Locale**: This option enables you to choose your default language and corresponding character set. The Pi is configured for UK English by default, but if you want to make a change, you can do so here. For instance, users located in the United States should select the en_US.UTF-8 locale.

- **Change Timezone**: The Foundation cut some financial and production corners by not including a real-time clock on the Raspberry Pi PCB. Accordingly, the Pi needs some help in determining the current time and date. Please be sure to select your proper timezone here, and then as long as the Pi is connected to the Internet, the Pi will periodically synchronize system time with one of the world's atomic time clock servers.

- **Keyboard Layout**: By default, Raspbian is configured for the UK English keyboard layout. This makes sense because Raspberry Pi is a UK product. However, you'll want to change the keyboard layout to match your locale so that you don't see any unexpected behavior in your typing results. Most commonly, this "unexpected behavior" manifests in, for instance, a US user typing @ and seeing " (double quote) instead, or typing # and instead seeing the pound sterling symbol.

NOTE: INTERNATIONALISATION?

Given the Raspberry Pi is a product of the UK, if you are an American who is accustomed to using Zs instead of Ss ("internationalization" versus "internationalisation"), you should just jolly well get with the global program, wot?

Enable Camera

This option loads the Raspberry Pi camera module drivers and packages, enabling you to make use of the camera board. We'll spend a lot of time with the Raspi camera in Chapter 16, "Raspberry Pi Portable Webcam."

Add to Rastrack

This option enables you to add your Raspberry Pi to the Rastrack (http://is.gd/sGStJL) database. Rastrack is a live map that shows you the geographic distribution of Raspberry Pi computers. It's really cool—check it out!

We'll actually learn how to use Rastrack and address any privacy-related concerns you may have in Chapter 17, "Raspberry Pi Security and Privacy Device."

Overclocking

Overclocking refers to tweaking the CPU operating parameters to force the processor to run at a higher speed than it was originally designed for. The Foundation provides us with helpful overclocking levels to afford us the opportunity to turbo-charge our Pi while at the same time reducing the possibility of frying the chip. Again, this option is discussed in great detail in Chapter 18, "Raspberry Pi Overclocking."

In the meantime, let me at least let the proverbial cat out of the bag by showing you the overclock levels in Figure 5.9. I'm sure the wording in the dialog "Chose overclock preset" will be fixed in a future firmware update (at least I hope so; sometimes the Pi reveals the lack of spit and polish inherent in grassroots community projects).

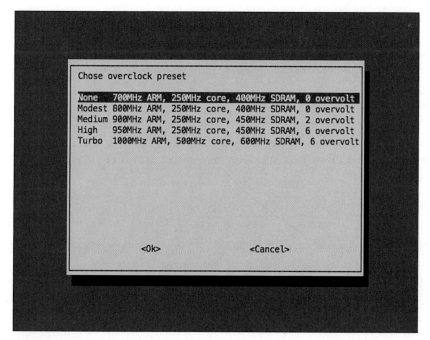

FIGURE 5.9   The Foundation makes it pretty easy to overclock the Raspberry Pi CPU.

Advanced Options

This is a submenu that contains the following options:

- **Overscan**: The overscan option enables you to manually adjust the Raspbian screen image. I've never had a problem with the display spilling off the outer border of my monitor, but it's nice to have correction capability built into the OS.

- **Hostname**: This option enables you to change the name of your Raspberry Pi from its default name of raspberrypi to something of your choosing. I've found this option helpful when I'm dealing with a busy network that consists of more than one Raspberry Pi.

  Changing any default value in a computer is a recommended security practice because an attacker's first task is to breach your security by capitalizing upon defaults that were never changed by the owners.

- **Memory Split**: The Pi's SoC consists of two processing centers: the CPU and the GPU. You can adjust how much memory should be reserved for the GPU by accessing this option. This can have a great impact on system performance depending upon what application you use. This option is also discussed in great detail in Chapter 18.

- **SSH**: Secure Shell (SSH) provides a secure and reliable means of establishing a command prompt session on a remote computer. The SSH server functionality is enabled in Raspbian by default, so you should not need to do anything here.

  You learn all about SSH connectivity in Chapter 7.

- **Update**: This option performs an update check for the Raspi-Config script itself.

- **About raspi-config**: This is purely an informational dialog.

Raspi-Config Under the Hood

If you'd like to view the Raspi-Config script source for intellectual curiosity's sake, run the following Terminal command:

```
nano /usr/bin/raspi-config
```

From inspection of the previous command you can draw the following conclusions:

- Raspi-Config is actually a Bash shell script (Linux script files typically have the file extension .sh).

- The Raspi-Config.sh script is located in the Raspbian file system in the /usr/bin directory.

Just for grins, I show you the partial contents of Raspi-Config.sh in Figure 5.10.

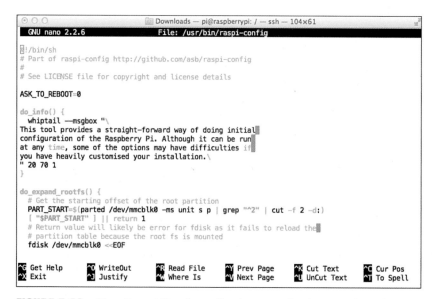

FIGURE 5.10 The Raspi-Config utility is actually the product of a Bash shell script.

Please note that as the Raspberry Pi Foundation adds new features to the Pi, they accordingly update the Raspi-Config utility. Therefore, don't be surprised if, after running an apt-get upgrade, you see a new Raspi-Config option or a slightly changed menu organization.

Next Steps

At this point I hope you now have enough familiarity with the Linux command-line environment and Raspi-Config utility that you can set up your Raspberry Pi and get into and out of the graphical shell.

In the next chapter, I formally introduce the LXDE graphical environment; doing so should answer some questions that probably popped into your head regarding this interface.

## CHAPTER 6

Debian Linux Fundamentals—Graphical User Interface

"Why run a graphical user interface (GUI, pronounced *gooey*) at all on a Raspberry Pi?" some computer enthusiasts complain. They'll tell you that the Pi's hardware is slow enough as it is. Not to mention that any self-respecting Linux user should be able to accomplish everything he needs by using the command line.

Yes, some Linux aficionados take user experience (UX) matters that seriously. And with the Raspberry Pi they have a point to a certain extent. Shouldn't we conserve as much of the Pi's CPU and GPU resources for actual work instead of for drawing fancy windows?

Well, friends, the raw truth of the matter is that many computer users simply prefer the simplicity and intuitiveness of a GUI. Most of us who have grown accustomed to the keyboard and mouse-based navigational methods in OSs such as Microsoft Windows and OS X don't want to bother learning something new. Thus, the Foundation saw fit to include a graphical shell in the Raspbian Linux distribution. In this chapter you learn how to start, configure, use, and exit the GUI shell.

LXDE—The Desktop Environment

The official Raspbian Linux distribution includes the Lightweight X11 Desktop Environment, or simply LXDE. The main benefit of LXDE is that it was written to be as light on system resources as possible, while at the same time offering the Linux user as much functionality as possible. I show you the default LXDE Desktop in Raspberry Pi later on in this chapter, in Figure 6.1. Ignore the annotations for now; the "Touring the LXDE Interface" section later in this chapter describes all of the user interface parts.

Oh, in case you are wondering—there isn't anything wrong with your video if you compared Figure 6.1 with your Pi's desktop and thought, "Why don't I see the blue background like Tim has?" I simply swapped out the default white background to one that isn't so harsh on my eyes. Take heart—you'll learn how to customize your LXDE environment soon enough.

Understand that LXDE is by no means the only game in town when it comes to Linux GUI desktops in general and system resource-friendly systems in particular. Some popular Linux GUIs include

- **KDE, GNOME** (pronounced *guh-nome*): These are the most popular GUI desktops on full Linux installations. However, you will be hard-pressed to run these GUIs on the Pi because they are too resource-intensive.
- **XFCE**: This is the main competitor of LXDE in terms of low-overhead Linux.

NOTE: X FACTOR

X-Windows, X Window System, and X11 are all synonyms that refer to the software system and network protocol that makes GUI desktop environments available locally or remotely on Linux systems. You can consider X the "engine" that powers the actual Windowing systems such as GNOME, KDE, or LXDE. Many Linux fans use X generically to refer to the GUI.

Starting and Exiting LXDE

If you started your Pi the traditional way, which is to say, by booting to the terminal prompt, then you can start LXDE by typing **startx** and pressing Enter on your keyboard.

The most straightforward way to close your LXDE session and return to a shell prompt is to click the Logout button on the LXPanel task bar. We'll delve deeply into LXPanel as we move through the chapter.

Remember our good friend raspi-config? We can customize boot behavior using this tool.

TASK: USING RASPI-CONFIG TO ADJUST BOOT BEHAVIOR

To configure your Raspberry Pi to boot directly to the graphical desktop, follow these steps:

1. From a Raspberry Pi terminal prompt, type **sudo raspi-config**.

2. From the Raspi-Config main menu, arrow down to the **boot_behaviour** option and press Enter.

3. Answer the question *Should we boot straight to desktop?* by typing either **Yes** (to enable GUI boot at launch) or **No** (to disable GUI autolaunch).

4. When you exit Raspi-Config, answer the question *Would you like to reboot now?* by typing in **Yes** and pressing Enter.

Touring the LXDE Interface

Using Figure 6.1 as our guide, let's review each of the Desktop icons and user interface elements in LXDE. Please continue to ignore the annotations at the bottom of the figure for now; we'll cover those in just a moment. For now concentrate instead on the icons that run along the left-hand side of the screen.

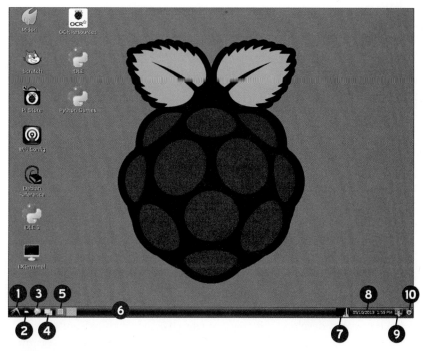

FIGURE 6.1   LXDE is the default GUI shell in Raspbian.

- **Midori**: This is a shortcut to Midori, which is (surprise-surprise) a minimalist web browser that conserves system resources.
- **Scratch**: This is a link to Scratch 1.4. You learn how to create Scratch programs in Chapters 8, "Programming Raspberry Pi with Scratch—Beginnings," and 9, "Programming Raspberry Pi with Scratch—Next Steps."
- **Pi Store**: This is a marketplace for the exchange of Pi-compatible apps. The Pi Store later is covered in more detail later.
- **WiFi Config**: This is an incredibly helpful tool for setting up a WiFi connection. You learn about the Pi's networking options in Chapter 7, "Networking Raspberry Pi."
- **Debian Reference**: This is a shortcut that opens the Dillo web browser and presents the Debian reference pages in HTML format. These files are stored locally, so you don't need to be connected to the Internet to read the pages. You can see this interface in Figure 6.2.

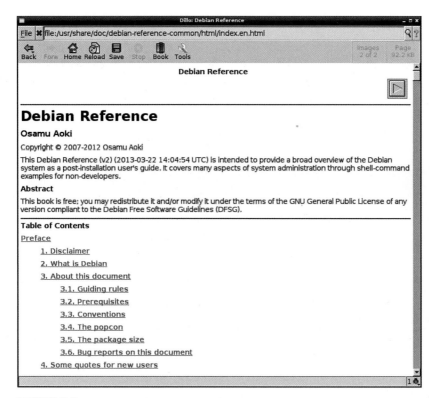

FIGURE 6.2   Raspbian gives you offline access to Debian Linux help files.

- **IDLE 3**: This is the integrated development environment (IDE) for Python 3. As you learn in the next chapter, Raspbian includes both Python 2 and Python 3. Python is the default and recommended programming language for the Raspberry Pi (recall that the "Pi" in "Raspberry Pi" is an indirect allusion to the Python language).

- **LXTerminal**: This is the command-line interface for the Pi. You can do everything in LXTerminal that you can from outside of X in the terminal environment.

- **OCR Resources**: This is a shortcut that opens Midori and opens the ICT and Computing page of the Oxford Cambridge and RSA Examinations (OCR) website. Basically, this is a site for teachers to get ideas on how to integrate Raspberry Pi into their academic curricula.

- **IDLE**: This is the IDE for Python 2.

- **Python Games**: This is a simple app launcher that enables you to try out a handful of Python games.

Now let's turn our attention to the annotations at the bottom of Figure 6.1. Because there are so many tiny icons, I thought than numbering them would make it easier for you to differentiate them.

1. **Main Menu**: An app launcher that functions the same way the Windows 7 Start menu does.

2. **PCManFM**: A lightweight file manager that functions the same as Windows Explorer or the OS X Finder. You can take a look at PCManFM in Figure 6.3.

FIGURE 6.3 The PCManFM file manager.

3. **Midori**: The lightweight web browser.

4. **Minimize All**: When you click this button, all onscreen windows minimize to the LXPanel.

5. **Virtual Desktops**: Linux has had this capability for a long time. You get two desktops by default, but you can customize this number by right-clicking one of the virtual desktop icons, selecting Desktop Pager Settings from the shortcut menu, and adjusting the Desktops options in Openbox Configuration Manager.

NOTE: UNDERSTANDING VIRTUAL DESKTOPS

A virtual desktop in the context of Linux in general and the Raspberry Pi in particular is a software-based method for extending your computer's desktop environment beyond the borders of the physical monitor. That is, you can have multiple copies of the desktop that each display separate application or document windows. This technology is less expensive, although less convenient, than connecting multiple physical monitors to your computer.

6. **LXPanel** (the bottom bar as a whole): Behaves exactly like the Task Bar in Windows 7. Actually, every element you see on this bar is nothing more than an applet that is enabled in LXPanel. To customize LXPanel, right-click an empty part of the task bar and select Panel Settings. Select the Panel Settings option to turn LXPanel panel applets on or off.

7. **CPU Usage Monitor**: Gives you real time, at-a-glance data on how busy the Pi's CPU is.

8. **Digital Clock**: Right-click this applet and select Digital Clock settings to edit it. The clock format uses a wonky, propeller-head style; visit http://is.gd/XHcNwN to learn the syntax.

9. **Screenlock**: You click this button to protect your Pi against unauthorized access while you aren't using the system.

10. **Logout**: You use this control to leave X and return to a terminal prompt, reboot the system, or shutdown (halt) the Pi.

Buttons B–D represent what is called the *Application Launch Bar*. The three shortcuts that exist there by default are simply that—defaults. To add additional shortcuts to the Application Launch Bar, right-click one of the app shortcuts and select Application Launch Bar Settings from the shortcut menu.

Essentially the entire bottom task bar in LXDE represents the modular LXPanel. It's quite a nifty toolset, actually. For more information on LXPanel, visit http://is.gd/yINVrP.

Delving into the Main Menu

The Start menu has worked well for Microsoft operating systems since Windows 95. Personally, I'm furious that Microsoft removed the Start menu from Windows 8; however, that is a discussion for another time. The good news for fans of the functionality the Windows Start menu used to give us, is that you'll find one very much like it here.

Because the LXPanel Main Menu is such a central part of the LXDE user experience, I wanted to spend a bit of time giving you the lay of the land with this tool. You can see the expanded Main Menu in Figure 6.4 (I highlighted the Main Menu button to make it easier for you to identify).

FIGURE 6.4 The LXPanel Main Menu functions similarly to the Windows 7 Start menu.

Following is the top-level navigation of the Main Menu:

- **Accessories**: Here you will find links to an image viewer, the Leafpad text editor, LXTerminal, and the Xarchiver zip/unzip tool, among other tools.

- **Education**: Here you can launch Scratch or its underlying Squeak management tool. Scratch is based on an old programming language called Smalltalk, and Squeak is the specific Smalltalk dialect that is used in Scratch.

- **Graphics**: Here you can view Adobe Acrobat Portable Document Format (PDF) files by using the open-source xpdf viewer.

- **Internet**: Here you can open the Midori and Dillo lightweight web browsers, as well as configure Wi-Fi settings.

- **Office**: Here you find the tiny yet functional Orage Calendar and Orage Globaltime utilities. These tools are actually part of the XFCE desktop environment briefly discussed earlier in the chapter.

- **Other**: This is a catch-all bucket for tools that the Raspbian developers could not or did not want to pigeonhole into another program group. You'll find a little of everything in this folder.

- **Programming**: As expected, from here you can launch Scratch or the Python 2/3 IDLE development environments.

- **Sound & Video**: Here you can open a simple audio mixer application to configure Pi's onboard audio. Remember that system audio comes directly from the Broadcom SoC; there is no dedicated sound "chip" on the Pi PCB.

- **System Tools**: Here you can rename files, view and manage running tasks, and use the Thunar File Manager (Thunar is nothing more than an alternative to PCManFM).

- **Preferences**: This folder contains several file and system management utilities.
- **Run**: You can use Run to issue terminal prompt statements from the GUI. This tool is directly analogous to the Run box in Windows 7.
- **Logout**: Exit and return to the command prompt environment.

The ordering and organization of the LX Panel Main Menu feels haphazard and not well thought out to me. Surprisingly (or not, depending on your attitude about Linux), there is no graphical method for customizing the Main Menu. The LXDE Wiki shows you how to manage the *.desktop files in the /usr/share/applications directory to customize the Main Menu; see http://is.gd/OTcpFy for more details.

Installing, Updating, and Removing Software

In Chapter 5, "Debian Linux Fundamentals—Terminal," you learned how to use the apt-get commands to update configured software repositories and update all installed software, including Raspbian itself. Open up LXTerminal, and let's now revisit the Advanced Packaging Tool, or *apt*.

First, what is a repository? In Linux terminology, a *repository* is an online source of regularly updated installation packages. Repositories are specific (almost) to every Linux distribution.

From LXTerminal, type **sudo nano /etc/apt/sources.list** to view the list of Raspbian repositories. On my system, the file contained only one line:

```
deb http://mirrordirector.raspbian.org/raspbian/ wheezy main contrib non-free rpi
```

Type **CTRL+X, N** to exit the sources.list configuration file without saving any changes.

Now let's review how to find and install software by using apt. But first let's update the repository:

```
sudo apt-get update
```

If you know the name of the software package that you are interested in (let's say the XFCE graphical environment—why not?), run this command:

```
sudo apt-get install xfce4
```

If you don't remember the precise name of the package of interest, try this:

```
sudo apt-cache search xfce4
```

To get a list of installed software packages, you should switch gears just a little bit and use the dpkg command instead, like so:

```
sudo dpkg —get-selections > ~/Desktop/packages
```

What the dpkg command does in this context is generate a list of installed packages into a file called packages that will appear on the LXDE desktop. Remember that the tilde (~) character denotes your home directory.

You can then view the list of installed packages by running

```
nano ~/Desktop/packages
```

You should know that the apt tools we've been using so far represent a (more) user-friendly front-end to the dpkg package management system that has been a component of Debian Linux for quite some time.

In Chapter 5 you were introduced to the man command. Be sure to access the man pages for the apt-get and dpkg commands so you can learn more about them.

To uninstall an installed software package, try this:

```
sudo apt-get remove xfce4
```

NOTE: PERFORMING A COMPLETE UNINSTALL

To remove both an installed application package as well as any associated configuration files, run the command **sudo apt-get —purge remove pkgname**.

Finally, by way of review: You'll want to run the following command periodically to update the repositories known by your Pi system:

```
sudo apt-get update
```

Remember that running update is only half the battle; you must also install any detected updates:

```
sudo apt-get upgrade
```

Accessing the Pi Store

The Pi Store, shown in Figure 6.5, is the Raspberry Foundation's version of the Microsoft Store or the Apple Store. The Store is meant to be a place where you can easily find software that has been developed specifically for the Raspberry Pi platform. You can start the Pi Store by double-clicking the desktop shortcut.

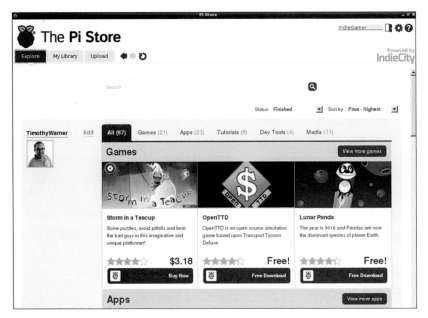

FIGURE 6.5   The Pi Store is to the Raspberry Pi what the Apple Store is to a Mac.

To do anything meaningful at the Pi Store (this means downloading software or uploading your own stuff), you must first create a free account. To do this, click Login in the upper-right corner of the Pi Store interface and then click Register.

You'll observe a few things about the basic operation of the Pi Store. One, as of spring 2013, there aren't a whole lot of apps in the Store. This situation should improve over time as the Pi develops a critical mass and sizable user base. Two, apps are of both the free and paid varieties. Also you need a PayPal account (http://www.paypal.com) to purchase a Pi Store app.

As with everything else in the open source community, the Pi Store is all about sharing your work with others. Visit the Upload tab in the Pi Store to register as a developer and then begin the process of building and sharing your Pi Store games and apps. There is even the possibility of some extra money if your paid app is successful! Figure 6.6 shows you my newly created developer page.

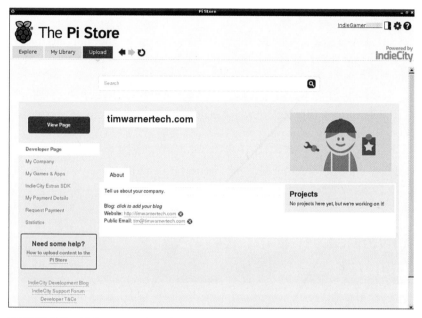

FIGURE 6.6   When you register as a developer, you get your own profile page in the Pi Store.

TASK: DOWNLOADING A FREE APP FROM THE PI STORE

1. From the LXDE desktop, double-click the Pi Store icon to open the program.

2. In the upper-right corner of the Pi Store, click Login, and then click Register to create a free user account.

3. Once you're logged in, navigate to the Games section and find the free app called CrazyWorms2. If for some reason the game is no longer available, choose another free game.

4. Click Free Download to download the game to your library. Note that the Pi Store takes you from the Explore tab to the My Library tab.

5. To play your newly downloaded game, select CrazyWorms2 in My Library and then click Launch in the right-hand information panel. Have fun!

Tweaking the LXDE UI

In this section I take a task-centered approach in showing you the most common interface tweaks in LXDE. It's important that I show you how this works so you can make the LXDE graphical environment your own.

For instance, I find the default white background color to be much too harsh for my light-sensitive eyes. Thus, the first interface tweak I always make on my Pi systems is to swap out the background color to something a bit more neutral.

I think you'll find that LXDE gives you the same kind of control you have in, say, Windows or OS X, but the interface controls are a bit less intuitive to find and use. (Hey, welcome to Linux!)

TASK: CHANGE THE DESKTOP BACKGROUND

Speaking of swapping out the desktop background, why don't we use that as our first procedure?

1. To tweak the LXDE desktop background, right-click an empty area of the desktop and select Desktop Preferences from the shortcut menu (see Figure 6.7).

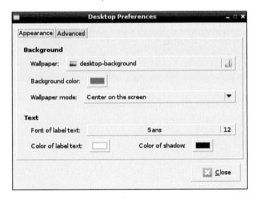

FIGURE 6.7 You can customize the LXDE desktop by using GUI configuration applets.

2. You can perform the following actions in the Desktop Preferences dialog box:
- Change the wallpaper independently of the background color. For instance, to specify a solid color background, set the Wallpaper Mode property to Fill with background color only and then adjust the Background color property.
- Adjust the font and color of icon label text. To do this, adjust the Font of label text, Color of label text, and Color of shadow properties.

TASK: CUSTOMIZE THE LXTERMINAL

I find that even when I'm working in a Linux GUI I still have at least one Terminal session going on at any point in time. Let's make that command prompt environment as user-friendly as possible, shall we?

1. Open LXTerminal and then click Edit, Preferences.

2. In the LXTerminal dialog box, make any necessary changes. My "tricked out" LXTerminal along with the Preferences window are shown in Figure 6.8. You can perform several terminal configuration tweaks:

- Change the terminal font and colors. The background color denotes the LXTerminal screen, and the foreground color refers to the onscreen text. You can do this by setting properties on the Style tab.

- Customize the cursor. For instance, you can change the cursor style from block to underline by editing the Cursor style property on the Style tab.

- Edit the default scrollback value by adjusting the Scrollback lines property on the Display tab. This is an important setting because it determines how far back you can scroll in an LXTerminal session. For instance, you may display a large text file onscreen and you want to ensure that you can scroll back to see every single line of text.

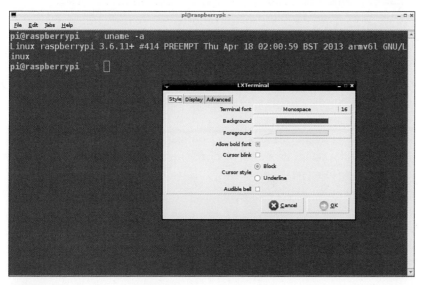

FIGURE 6.8 You can customize LXTerminal to suit your individual preferences.

TASK: CUSTOMIZE THE LXPANEL

LXPanel represents the primary navigation system used in LXDE, as you've already learned. Each component that you see on the Task Bar at the bottom of the LXDE screen represents a particular LXPanel applet. Follow these steps to visit the major configuration touch points:

1. Right-click an applet in the panel and select the first option in the shortcut menu to customize that particular applet. For instance, you can right-click an empty area of the Task Bar and select Task Bar (Window List) Settings to customize the position, appearance, and so on of the task bar. Right-clicking the CPU meter gives you the option CPU Usage Monitor Settings.

2. From any LXPanel shortcut menu, select Panel Settings to open the global configuration dialog for the panel. The Panel Preferences dialog box consists of four tabs:

   - **Geometry**: Position, size, alignment, and icon of the LXPanel bar.

   - **Appearance**: Background color and/or wallpaper used to "color" the LXPanel.

   - **Panel Applets**: Add, remove, and edit the applets that are included on the LXPanel. You can also specify the order in which you want these applets to appear.

   - **Advanced**: Set preferred applications, task bar hiding, and so on.

Openbox

Let's finish this part of the chapter by briefly discussing Openbox. Openbox is different from LXPanel; LXPanel is a navigation user interface host, but Openbox is the Window Manager that runs behind LXPanel.

Open the Main Menu and click Preferences, Openbox Configuration Manager. Here you can make further tweaks to applets resident in the LXPanel (the number of virtual desktops, for instance) and globally in the system (see Figure 6.9).

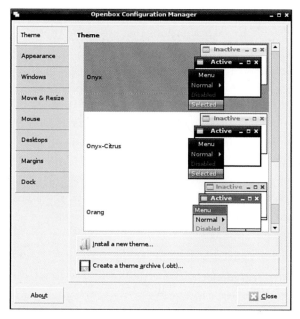

FIGURE 6.9   The Openbox Configuration Manager.

Actually, you can find pretty much everything you want or need, GUI look and feel-wise, by browsing the Preferences program group of the LXPanel Main Menu.

Editing Configuration Files

You'll find that despite the control that GUIs like LXDE give you over the environment, the unavoidable truth is that you accomplish most tweaking and system configuration by editing text files.

For instance, let's say that you want to disable the Raspbian screensaver (it looks cool, but it consumes valuable system resources). Now it is true that you can disable or enable the XScreensaver in LXDE by opening the Main Menu and clicking Other, Disable XScreenSaver or Other, Enable XScreenSaver, respectively. You can also change the currently active screensaver by opening the Main Menu and clicking Preferences, Screensaver. However, you can also permanently disable the screensaver by editing the LXDE autostart file with

```
sudo nano /etc/xdg/lxsession/LXDE/autostart
```

When you are in the file, you need to remove the following line from the file:

```
@xscreensaver -no-splash
```

Simply place your cursor at the end of the line and use Backspace to remove it. Next, press Ctrl+O to save changes and then press Ctrl+X to exit the nano editor. Done and done!

Remember always to invoke sudo when you attempt to edit a configuration file; you won't be able to save your changes due to the default access permissions that are set on those files.

You are probably asking, "Tim, how did you know to edit the /etc/xtg/lxsession/LXDE/ autostart file?" My answer is simple: "Google it." With Linux, we have a huge community of computer scientists, experts, and enthusiasts at our disposal. I like to tell my students, "If you are experiencing a problem with your Pi, chances are good that hundreds of other people are having the same problem, and somebody has figured out how to fix it."

To that end, let me share with you some of my favorite Raspberry Pi discussion forums. These online discussion boards are a wonderful way not only to get your questions answered, but also to connect and swap tips with other Raspberry Pi enthusiasts who are located throughout the world:

- **Official Raspberry Pi Forums**: http://is.gd/6nBR5Z
- **Element 14 Raspberry Pi Forums**: http://is.gd/2urLqa
- **Elinux.org Communities Reference**: http://is.gd/yleGIw

## CHAPTER 7

Networking Raspberry Pi

Basic Networking Concepts

Ethernet is the *de facto* networking standard nowadays. It's a protocol (set of protocols, actually) that represents an agreed-upon set of rules that computing devices use to establish digital communications with each other.

The good news is that any self-respecting local area network (LAN) and certainly the Internet all use Ethernet, so there really isn't anything more we need to discuss at that level.

More granularly, Ethernet hosts (that is to say, any device that has a network interface card [NIC] installed and is configured for Ethernet networking) must have a unique Internet Protocol (IP) address to be able to send and receive data meaningfully.

There are two versions of IP in use in the world today: IP version 4 (IPv4) and IP version 6 (IPv6). Because IPv4 is the current standard and the state of IPv6 remains somewhat in flux, this book focuses solely on IPv4.

An IPv4 address looks like this:

```
192.168.1.204
```

The IPv4 address is a "dotted decimal" representation of 32 binary digits, or bits. At base, you need to know that each of the four decimal numbers ranges from 0 to 255 and that each host on a network must have a unique IP address.

The other 32-bit number that is used in conjunction with the IP address is called the subnet mask. For instance, a typical subnet mask that is used on many home networks is

```
255.255.255.0
```

The "mask" in "subnet mask" is used to differentiate the shared network portion of the IP address from the computer-specific host portion. For instance, the IP address/subnet mask combination 192.168.1.204/255.255.255.0 means that this host resides on the 192.168.1.0 network, and the station's unique identifier is 204.

Actually, it is useful to compare the network and host portions of an IP address to a street address. If my mailing address is 110 Smith Street, then "Smith Street" represents the shared network portion (on which other houses reside), and "110" represents my unique host ID.

The Raspberry Pi will in all likelihood receive its IP address automatically from a router or dedicated server running the Dynamic Host Configuration Protocol (DHCP) service. For instance, in my home office my Comcast Business Gateway (a fancy term for a combo cable modem and router) leases IP addresses to all of the hosts in my network. This all happens automatically and, usually, without any need for me to get involved.

NOTE: MORE ON HOSTS

Remember that any device with an installed NIC that communicates on an Ethernet network is known as a host. This includes PCs, Macs, network printers, "smart" switches, game consoles, Internet-capable TVs and DVD players, mobile devices—the sky is almost the limit. And being TCP/IP hosts, each host needs some method for obtaining a unique IP address.

Thus, it is possible to assign specific IP addresses to your Internet-enabled devices or to use dynamic IP configuration. By the conclusion of this chapter, you'll understand how to use both IP addressing methods.

Configuring Wired Ethernet

If you have any interest in having your Raspberry Pi communicate with other hosts on your home network, or perhaps with Internet-based resources, then you need to know how to configure networking.

The Raspberry Pi (depending upon the revision, whether A or B), supports both wired and wireless Ethernet. To be a well-rounded Raspberry Pi power user, you should understand how each method works.

Let's start with wired Ethernet.

Before you put down the Raspberry Pi Model A for not having an RJ-45 port to support traditional wired Ethernet, remember the purpose of the Model A: to provide a stripped-down computer with a minimal power footprint. If you need Ethernet on the Model A, you can pop in a USB Wi-Fi dongle and go about networking that way. I talk about Wi-Fi more in a little while.

For those who have a Model B, it's time to plug in a standard Ethernet cable into the RJ-45 jack and power the device on. As I said earlier, the Pi should pick up an IP address that is valid for your network from a DHCP server. This DHCP server can be a wired router, a wireless router, or a dedicated server.

Take a look at Figure 7.1, which shows the TCP/IP configuration on my Pi. I use the ifconfig (pronounced eye-eff-config or ihf config) command for this purpose.

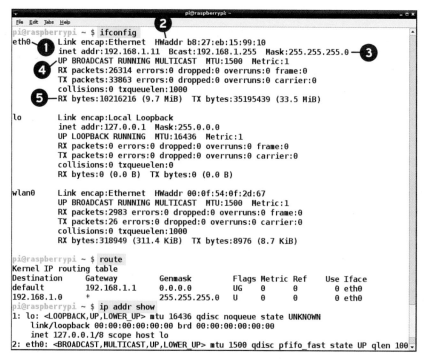

FIGURE 7.1 You can use ifconfig and route to view TCP/IP client configuration on your Raspberry Pi.

Using the annotations on Figure 7.1 as our guide, here's what all that output means:

1: These are network interface IDs. The eth0 interface refers to the RJ-45 wired Ethernet interface on the board. The lo interface represents the loopback interface, which is used for testing and diagnostics. If you have a Wi-Fi interface, you'll see an entry called wlan0.

2: The **HWaddr** refers to the network interface's media access control (MAC) or hardware address. This is a unique identifier that is permanently "burned" into the network interface by the manufacturer.

3: The **Mask**, or subnet mask, is a string of binary zeros (255 when translated into decimal notation) that serves to separate the network portion from the host (unique) portion of an IP address. If you didn't know that an IP address consists of two (and often three) parts, then I guess you just learned something new!

4: The **Up** or **Down** status notifications are useful for troubleshooting purposes.

5: This is Send/Receive metadata that is most useful when tuning network performance or undergoing diagnostics.

Besides the IP address proper and subnet mask, another important IP address you should know is the *default gateway*. This is the IP address of your router; the router is the device that gets your Pi out from your local area network (LAN) to the Internet.

You can view your Pi's current default gateway by issuing the route command from a Terminal session. You can see the route command output in Figure 7.1 as well; specifically, look for the IP address under the Gateway column.

NOTE: THE HEART OF PI'S WIRED NETWORKING

Be careful to differentiate the RJ-45 "ice cube" port on your Raspberry Pi and the actual network interface circuitry. The Ethernet engine on the Model B board is the LAN9512 IC that is located on the PCB directly behind the USB port stack.

Another way to view your Pi's TCP/IP configuration information is to issue the command **ip addr show**. The ip command is pretty robust; run man ip to view the man page.

Let's set a static IP address for the wired Ethernet interface on the Pi just for grins.

TASK: SETTING A STATIC IP ADDRESS ON YOUR RASPBERRY PI

To set a static IP address, you need to edit the interfaces configuration file.

NOTE: NANO NOT REQUIRED

Please know that you are perfectly free to use any Linux text editor when you edit configuration files. The only reason I use the nano editor in this book is because it is my personal favorite. Your mileage might vary—for instance, you could have a strong preference for vi. It's all good!

1. Run **sudo nano /etc/network/interfaces** to open the interface's configuration file for editing.

2. Change the line that reads:

```
iface eth0 inet dhcp
```

to

```
iface eth0 inet static
```

3. Below the changed line, add the following lines; in this example I am supplying "dummy" data just to show you what a typical configuration looks like:

```
address 192.168.1.100
netmask 255.255.255.0
network 192.168.1.0
broadcast 192.168.1.255
gateway 192.168.1.1
```

A comprehensive discussion of network addresses, broadcast addresses, and default gateways is well outside the scope of this book. Suffice it to say that you need to have at least a good solid knowledge base in networking before you undertake static IP configuration.

One more thing before you switch gears from wired Ethernet to wireless Ethernet—remember that the Pi board contains status LEDs. Pay particular attention to the FDX, LNK, and 100 lights. The FDX and 100 LEDs should glow solid, and the LNK light should flash as data is sent from and received by the Ethernet interface.

Configuring Wireless Ethernet

The wired Ethernet capability of the Model B board is all well and good. However, what if your Raspberry Pi project won't work with a network cable? For instance, what if you want to mount a security camera in your driveway? Do you really want to run Ethernet cable from your router out to the Pi? I don't think so.

Thus, you can configure wireless Ethernet (Wi-Fi) for both the Model A and Model B boards. You can find tiny USB Wi-Fi dongles all over the Internet; I recommend purchasing yours from Adafruit (I own this dongle, and it works great). You can see what it looks like in Figure 7.2.

FIGURE 7.2   Adafruit sells very reasonably priced USB Wi-Fi dongles. Regarding the annotations: 1 shows the Raspberry Pi in a nice case; 2 shows the two USB ports on the Model B board; 3 shows my Adafruit Wi-Fi dongle.

1: Raspberry Pi

2: USB ports

3: Wi-Fi dongle

NOTE: WHAT IS A DONGLE?

A *dongle* is a small piece of hardware that plugs directly into a computer, usually via USB. The dongle typically provides either copy protection for software or access to Wi-Fi networks.

Typically, configuring Wi-Fi under Linux is a nightmare. The good news is that the Raspberry Pi Foundation knows all about this problem and includes the wonderful WiFi Config utility for Raspbian. Without any further ado, let's set up Wi-Fi!

TASK: SETTING UP WI-FI ON YOUR PI

Even though the Raspberry Pi Foundation has done their best to simplify Wi-Fi setup, I think it best to walk you through the procedure step-by-step.

1. Turn off your Pi and plug your Wi-Fi dongle into your powered USB hub or Pi board's USB port. Although USB is technically a hot-pluggable technology, which means that you should be able to plug and unplug your dongle at will, Pi enthusiasts (myself included) have had problems with that. Thus, your best bet is to plug in the dongle prior to starting up the Pi.

2. Boot the Pi and launch the LXDE desktop.

3. Double-click WiFi Config, the interface for which is shown in Figure 7.3. Click Scan and then click Scan again in the Scan results dialog box. Find your preferred Wi-Fi network (the network must be configured to broadcast its SSID, unfortunately) and double-click it to specify your authentication and encryption options.

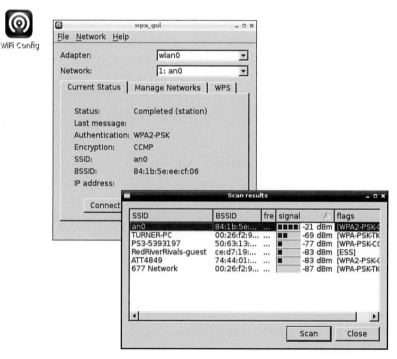

FIGURE 7.3 You can set up Wi-Fi easily by using the WiFi Config utility.

4. If your target Wi-Fi network is secured (and I certainly hope that it is), then fill in the necessary encryption and key parameters. Click Add to complete the configuration.

5. In the wpa-gui dialog box, ensure that your Wi-Fi network appears in the Network: field and then click Connect. You are now online with Wi-Fi!

After you are connected, you can check your Wi-Fi status by right-clicking its icon in the LXPanel Application Launch Bar (see Figure 7.4).

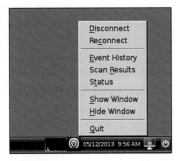

FIGURE 7.4   The WiFi Config utility runs in the LXPanel app launcher for easy access.

Configuring "Headless" Raspberry Pi

I've used the term "headless" Raspberry Pi a few times in the book so far. Just what the heck do I mean? Well, here's the deal: Computer monitors take up quite a bit of desk space. For instance, I have six monitors—big ones, too—in my home office! I don't want to stand up a seventh monitor to fire up my Pi.

"Headless" simply means that you connect to the Pi remotely without the necessity of an external monitor or television screen. You can use a couple networking protocols to make the remote access happen:

- **Secure Shell (SSH)**: This protocol gives you secure (encrypted) remote access to the Pi command prompt.

- **Virtual Networking Computing (VNC)**: This protocol gives you unsecure (unencrypted) GUI remote access to your Pi.

Before we get to using SSH and VNC, here's an initial prerequisite: You must have the Pi's IP address. How can you obtain this address, though, when you don't have a monitor available?

Your best bet is to download a freeware or shareware IP scanning tool. For Windows, I recommend the Advanced IP Scanner (http://is.gd/9qC1AI). This tool, which is shown in Figure 7.5, is extremely easy to use.

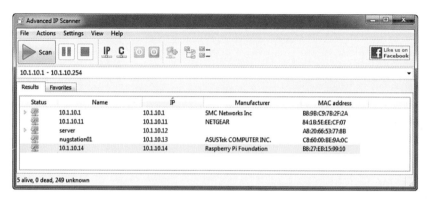

FIGURE 7.5 You can use an IP scanner to determine your Raspberry Pi's IP address.

The Advanced IP Scanner Tool parses the firmware metadata of any detected network interfaces; thus you can easily spot the RPi by looking for the entry with the Manufacturer entry of Raspberry Pi Foundation as shown in Figure 7.5.

On Apple OS X systems (and Windows and Linux computers as well, for that matter), I recommend Nmap (nmap.org). The Nmap toolset does a whole lot more than simply scan IP addresses; you can actually perform a lot of information security tasks with these programs.

Nmap can be run either from a command line or by using the built-in Zenmap graphical interface (the latter is shown in Figure 7.6). Note in the figure that Zenmap shows us the Pi both as a Linux box (which it is) and the Manufacturer field of the Pi's network interface.

Now let's turn our attention to how you can actually use the SSH and VNC protocols to remotely connect to your Raspberry Pi.

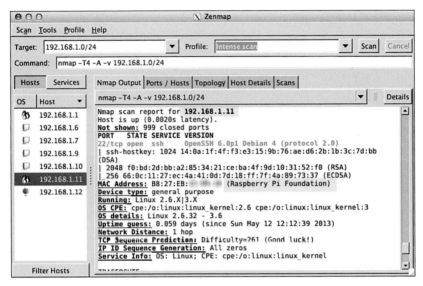

FIGURE 7.6 The Nmap/Zenmap toolset is a comprehensive suite of TCP/IP utilities for Windows, OS X, and Linux.

Secure Shell (SSH)

Secure Shell (SSH) is a Unix/Linux client/server network protocol you can leverage with the Raspberry Pi to support secure command-line remote access. Raspbian includes an SSH server and enables it by default. You can verify the SSH server status by opening Raspi-Config and checking the ssh (Enable or disable ssh server) option.

So the Pi is already set up as an SSH server. Now to establish a remote connection to the headless RPi, you need to use an SSH client. Unfortunately, Microsoft has never included SSH software in their operating systems.

Most people use PuTTY for Windows (http://is.gd/ResYA2), which you can check out in Figure 7.7. PuTTY is really simple to use; just fire up the tool, pop in the Pi's IP address in the Host Name field, and click Open.

FIGURE 7.7 On Windows systems, PuTTY presents a low-overhead way of connecting remotely to a "headless" Raspberry Pi.

TASK: USING SSH TO CONNECT TO A RASPBERRY PI REMOTELY

On OS X or Linux systems, you already have the SSH client built into the OS. Thus, you can fire up a Terminal session on your client system and perform the following procedure:

1. Assuming you are using the default user credentials of pi/raspberry and (in this example) the Pi is located at 192.168.1.11, you can issue the following command from the OS X or Linux terminal:

```
ssh pi@192.168.1.11
```

The previous command, when translated into conversational English, says that you want to use the SSH protocol to establish a remote connection to the SSH server listening at 192.168.1.11 and that you want to connect using the user account called pi.

2. Authenticate by providing the password for your local OS X or Linux account. You'll then be asked to verify the authenticity of the Pi. Because we know that this is the correct box, you can type **yes** and then press Enter to add the SSH server's public key to your system and automatically add the RSA thumbprint of the Pi to your /etc/.ssh/known_hosts configuration file.

3. Issue any RPi-specific commands (such as sudo raspi-config) to convince yourself that you are in fact remotely connected to your Pi. The entire SSH connection workflow from the perspective of OS X is shown in Figure 7.8. In the figure I highlighted the commands I used.

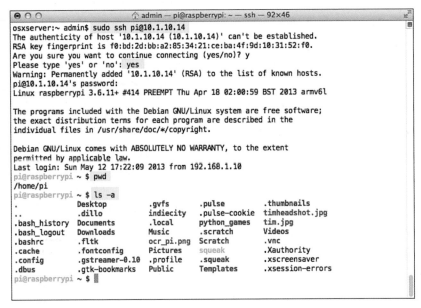

FIGURE 7.8   The SSH connection process to Raspberry Pi is straightforward.

The chief advantages to SSH-based remote access to the Pi are

- The client-side setup is quick and easy.
- You have full Terminal access to your Pi up to any restrictions that might be set on your connecting user account.
- All data transmitted between your remote workstation and the Pi is encrypted.

On the other hand, SSH remote connections to the Pi have one chief downfall—no GUI access. If you need to display an X Server desktop remotely, you need to turn your attention to setting up VNC.

Virtual Network Computing (VNC)

VNC is a high-performance and convenient method for sharing GUI desktops across a network. The two downsides to using VNC for our purposes are as follows:

- **By default, VNC transmits all data between the client and the server in plain text**. Therefore, if you have need for data confidentiality, you need to select an appropriate VNC server and client software package.

- **By default, Raspbian does not include a VNC server**. I show you how to address this issue immediately, so don't be overly concerned.

The VNC setup workflow consists of three steps: (1) installing a VNC server on the Pi; (2) installing a VNC client on your remote system; and (3) making the remote connection. Let's do this!

TASK: USING VNC TO CONNECT TO THE RASPBERRY PI

You will in all likelihood bookmark this page because using VNC to connect to your Pi is a procedure that you'll use on a regular basis with your Raspberry Pi. I'm glad to help!

1. On your Pi, fire up a Terminal session and run the following command to download and install the TightVNCServer (http://is.gd/A6k1nD). There exist many different VNC packages; Tight is simply considered to be a good choice for the Pi.

```
sudo apt-get install tightvncserver
```

2. Now you need to start the VNC server. You'll need to do this every time you boot the Pi unless you take steps to autorun the command. (You will learn how to set up the VNC server to run automatically in the next procedure.)

```
tightvncserver
```

3. Now it's time to start a VNC session. Again, you must do this manually every time the Pi is started unless you configure autolaunch:

```
vncserver :2 -geometry 1024x768 -depth 24
```

Here's what each part of the preceding syntax means:
- **vncserver :2**: This launches the VNC server session process and labels the session 2. Session 1 is started when you start the server. You can create additional sessions with different resolutions if you want. For instance, if you want the session to run in HD, try vncserver :3 -geometry 1920x1080 -depth 24.
- **geometry**: This determines the pixel size of the session window. 1024x768 is standard 4:3 aspect ratio, and 1920x1080 is 16:9 widescreen HD aspect ratio.
- **depth**: This number refers to the color bit depth for the VNC session. Twenty-four bits is the standard for the Pi.

Before you can test access from a remote workstation, you need to install a VNC viewer software. Again, lots of options exist here; I enjoy RealVNC (http://is.gd/EB07wO).

Start your VNC Viewer and specify the IP address and session number of your Pi. For instance, if my Pi listens for connections at 192.168.1.11 and I need VNC session number 3, I type:

```
192.168.1.11:3
```

Note there is no space between the IP address and the session number. You can see this in action in Figure 7.9.

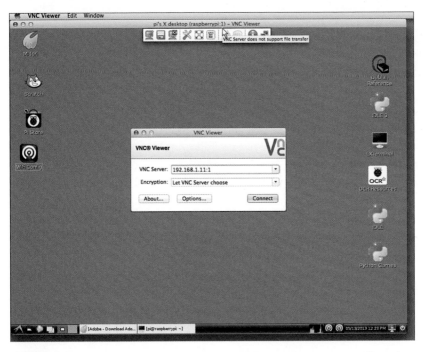

FIGURE 7.9   VNC gives you a remote GUI session on your Raspberry Pi.

Let's finish this section with a quick procedure on configuring your Pi to run the VNC server automatically at startup.

TASK: CONFIGURE YOUR PI TO START VNC SERVER AUTOMATICALLY

You probably don't want to run the steps in the preceding task every time you boot up your Raspberry Pi. Therefore, let me show you how easy it is to configure VNC to start automatically during every system startup.

1. Start Raspi-Config and ensure that you set the GUI to start automatically at every startup. This is done by navigating to the Enable Boot to Desktop menu option and answering Yes to the question *Should we boot straight to desktop?*

2. After the Pi comes back from its reboot, issue the following Terminal command to change your focus to the /home/pi/.config directory. Any file or directory with a period (.) in front of it means that it is hidden from view by default.

```
cd ~/.config
```

3. Create a directory called autostart. Reasonably enough, you do this in Linux with the mkdir command.

```
mkdir autostart
```

4. Let's now move the focus inside the new autostart directory.

```
cd autostart
```

5. You're almost finished. Create a new configuration file named tightvnc.desktop.

```
sudo nano tightvnc.desktop
```

6. Add the following lines to the new, blank configuration file. You can see my copy of the file in Figure 7.10.

```
[Desktop Entry]
Type = Application
Name = TightVNC
Exec = vncserver :1
     StartupNotify = false
```

7. Type Ctrl+X and then Y to save your changes and exit nano. Reboot the Pi, and you're done:

```
sudo reboot
```

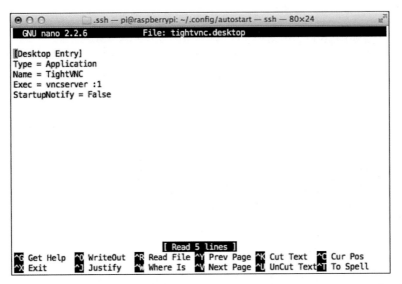

FIGURE 7.10 For some users, it is convenient to autostart the VNC server on the Raspberry Pi.

On Browsing the Web with the Pi

Midori (http://is.gd/5ccfPn) is the default web browser that the Foundation included in Raspbian (see Figure 7.11). I'm sure that the Foundation selected Midori because it is known as a "lightweight, fast, and free" web browser.

FIGURE 7.11   Midori is a good choice for the Raspberry Pi due to its low system resources footprint.

Here are some of the specific advantages of Midori as a Pi web browser:

- Highly adherent to web standards (although HTML5 support needs some work).
- Emphasis on security and user privacy.
- The Preferences panel allows you to suppress images and media from web pages, which improves browsing speed.
- Extensions support—extensions are browser add-ons that can greatly expand the capabilities of a web browser.

Frankly, the main limitations to Midori aren't really limitations of any web browser, but are more of a function of the Pi's own limited hardware resources and ARM processor architecture.

For instance, take Adobe Flash. Love it or hate it, there is much multimedia online that is viewable only if your browser supports the Adobe Flash Player plug-in. Unfortunately, Adobe abandoned Flash support for ARM processors quite a while ago. Therefore, out of the box, you can't view Flash (which includes YouTube) on the Pi. Bummer, right?

For hardcore Flash fans, you can try to hack around with the open source Gnash (http://is.gd/TvqqNI) player. You'll find that Gnash enables you to play Flash versions 7, 8, and 9 media objects, although you might be disappointed at the performance.

If you are as big of a fan of YouTube as I am, you'll be pleased to know that there are some Pi-specific options. Your best bet is to fire up your favorite search engine and perform a search for **play youtube raspberry pi** or something similar.

In Chapter 12, "Raspberry Pi Media Center," you learn how to build a Raspberry Pi media center by using the wonderful Xbox Media Center (XBMC) software.

Finally, if you tried Midori and simply concluded that you don't like it, you can certainly install an additional web browser. For the love of all that is holy, don't install a "full-sized" web browser like Mozilla Firefox on your Pi—you will live to regret it, I assure you.

Chromium (http://is.gd/oWiKFh), the open source fork of Google's Chrome browser, is a good choice for the Pi in my experience. Mozilla Firefox fans might want to take a look at Iceweasel (http://is.gd/cfmCHP); in fact, I'll be using Iceweasel as the default browser for the remainder of this book.

## CHAPTER 8

Programming Raspberry Pi with Scratch— Beginnings

Now that you've learned some of the history behind the Raspberry Pi, you understand why the Pi comes preloaded with Scratch, Python, and other software development environments. After all, the Pi's fundamental reason for being is to encourage schoolchildren to build an interest in computer programming.

Scratch is an intuitive programming language that was developed by the Lifelong Kindergarten group of the Media Lab at the Massachusetts Institute of Technology (MIT) in Cambridge, Massachusetts (http://is.gd/LYZJWm). The Media Lab folks wanted to present an easy-to-learn programming toolset that empowers interested kids (and adults) to quickly create games and other media-rich interactive experiences without having to understand complex syntax rules.

What's particularly cool about Scratch is its use of drag-and-drop *blocks* instead of traditional text commands to build computer programs. The MIT Media Lab wanted to help students avoid stumbling over the learning curve inherent in memorizing complex programming language syntax and instead focus on using logic and intuition to create Scratch projects.

So in a nutshell, Scratch is a way to teach nonprogrammers how to program. How cool is that?

To illustrate this point, take a look at Figure 8.1, which shows the same programmatic procedure (specifically an if/then condition) in C and in Scratch. Which syntax is more immediately relatable to you?

```
int testnum = 6
if (testnum > 6)
  printf("Greater than 6");
else
  printf("In range");
```

FIGURE 8.1 A conditional logic expression in C (top) and in Scratch (bottom).

Technical Aspects of Scratch

Scratch is an open source integrated development environment (IDE) that was itself built in Squeak, a dialect of the Smalltalk programming language. As you have probably come to expect with open source software, you can install Scratch on Windows, OS X, or Linux. Plans are already underway to port Scratch on mobile platforms. As I said earlier, the Raspberry Pi Foundation includes Scratch 1.4 in its Raspbian operating system distribution.

Because Scratch is a cross-platform application, you can create a .sb Scratch project file on your Pi, save the project file to an online storage service, download the .sb file to your PC, Mac, or Linux box, and resume your work seamlessly. Of course, this process works just as well in reverse.

Scratch Version Issues

I have good news and bad news for you. The good news is that Scratch 1.4 works just fine on the Raspberry Pi, and there are plenty of Scratch 1.4 programmers in the world. The bad news (if you want to call it that) is that the Lifelong Kindergarten Group released Scratch 2.0 in May 2013, and it is incompatible with the Raspberry Pi.

The incompatibility centers on the fact that Scratch 2.0 is built with Adobe Flash, and Flash is not supported on the Pi. By contrast, the Pi includes kernel support for Squeak, the underlying programming language behind Scratch 1.4.

I chatted with Eben Upton, co-creator of the Raspberry Pi, regarding the likelihood that the Foundation will find a way to bundle Scratch 2.0 on the Pi. This was his response:

We don't expect to support Scratch 2.0 any time soon, but we are doing a lot of work to make Scratch 1.4 more performant on the Pi (Smalltalk optimization and updating the underlying Squeak VM). In practice, we don't expect the Scratch team's use of Flash to last very long (even Adobe doesn't believe in that stuff anymore), so hopefully we'll reconverge when they port to JavaScript or similar.

So...the long and the short of it is that functionally Scratch 1.4 and Scratch 2.0 are practically identical, so everything you learn to do in this chapter is directly applicable to Scratch 2.0. That said, I focus educational efforts in this chapter strictly on Scratch 1.4 because that is the version included in Raspbian.

The Scratch Community

Another important principle behind Scratch is the power of community. Before going too much further with this discussion, I want you to visit the Scratch home page at scratch.mit. edu and register for a free Scratch user account (see Figure 8.2).

FIGURE 8.2 You should become a part of the Scratch community.

1: Browse Scratch projects here

2: Create a Scratch account here

Becoming a member of the Scratch community enables you to download projects made by other Scratchers (yes, Scratch users call themselves "Scratchers"), as well as to share your own work. Scratch site members can also post and answer questions on the Scratch discussion forums (http://is.gd/S4ZeId).

NOTE: CAT SCRATCH FEVER

The name *Scratch* does not officially arise from the feline world, although the Scratch Cat is the official mascot of the product. Instead, scratch in this context refers to the disc jockey (DJ) technique of scratching vinyl records on a turntable to produce interesting musical beats and variations. This notion of scratching is linked to the creativity and self-expression that is encouraged in building Scratch games.

The only downsides I've found to using Scratch on the Pi are as follows:

- Because the Pi does not support either Adobe Flash or Java, you can't preview Scratch games on the Scratch website.
- Because the Pi isn't exactly a robust computer, overall Scratch performance is a bit on the slow side.

If you want to install Scratch on one of your full-fledged desktop computers, it's easy to do. Scratch is a cross-platform application, which means that you can install and run it on Windows, OS X, or Linux. Download Scratch from the Scratch website (http://is.gd/iIuIQK).

One thing you will notice about Scratch 2.0 when you view the site is that you no longer need to download and install the environment. Instead, you can jump directly into the editor from the Scratch website.

Getting Comfortable with the Scratch Interface

To start Scratch, simply double-click the icon on your Raspbian desktop. An annotated version of the Scratch interface is shown in Figure 8.3.

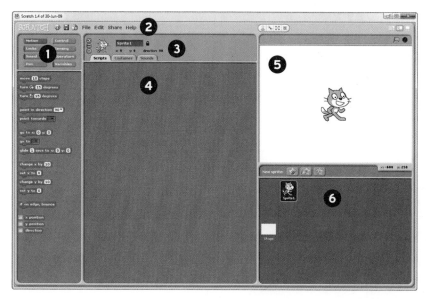

FIGURE 8.3   The Scratch user interface is friendly and intuitive.

1: Blocks Palette: You construct Scratch code by dragging and dropping action blocks.

2: Menu Bar: Here you can save a project, interact with the Scratch website, get online help, and perform other file management tasks.

3: Sprite Header Pane: This area displays important details for the currently selected sprite, including X/Y coordinates, name, and positional restrictions.

4: Scripts Area: This is where you actually program the logic of your Scratch application. Each sprite can have one or more script "stacks" associated with it. You'll notice that the Scripts area has three tabs:

- **Scripts**: For your code blocks.
- **Costumes**: Used to alter the appearance of your sprites (for the Stage, this tab is called Backgrounds).
- **Sounds**: Attach recorded or imported audio files to your sprites.

5: Stage: The Stage is the work area for our Scratch application. This is where all the action occurs. The Stage Toolbar, located above the Stage, gives you control over sprites and allows you to resize the Stage three different ways.

6: Sprites Pane: In case you wondered, a sprite is a graphical object that you include in your Scratch app. By default, the Scratch Cat appears as a sprite in all new Scratch projects. You can import existing graphics as sprites or draw your own from, well, scratch.

To get started, visit the Scratch Projects site (http://is.gd/tsr9gM) and download somebody's project that looks interesting. Again, you need to be logged in with your Scratch account to download projects (see Figure 8.4).

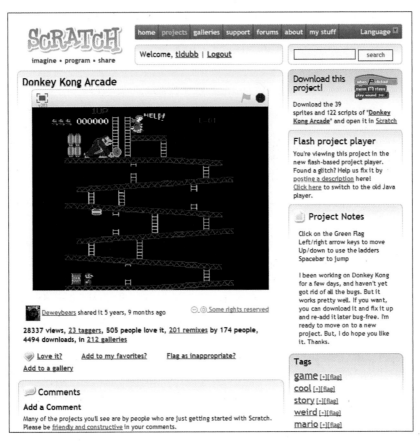

FIGURE 8.4 You can try out and download other peoples' Scratch projects directly from your web browser.

It bears repeating that you cannot preview projects in a browser on Raspi because the Pi does not support Adobe Flash. Moreover, my communication exchanges with Eben tell me that the Raspberry Pi Foundation has no future plans for the Pi to support Flash.

The main Scratch Projects website contains only Scratch 2.0 projects. Although you can upload your Scratch 1.4 projects to the website, they will be converted to Scratch 2.0 before they appear on the website. You can then edit the uploaded projects directly in your web browser (unfortunately, you can't run Scratch 2.0 projects from Scratch 1.4).

The traditional way to start a Scratch project (whether from a web browser or from within Scratch itself) is to click the Green Flag icon in the Stage area. By contrast, you can use the Red Stop Sign icon to manually stop the project.

One of the awesome things about open source software such as Scratch is that you can go beyond simply admiring other Scratchers' projects—you can actually view their source code and base your own Scratch projects off of that code.

Let's spend some time getting to know how the actual programming code works in Scratch. To that point, you need to understand what blocks are and how they are used.

About Blocks

As I said earlier, you can use these easy-to-understand blocks to actually program your Scratch app. As you'll learn soon enough, blocks are puzzle-type pieces that "snap" together in much the same way that LEGO blocks do.

These blocks make it easier for beginning programmers to think about and execute programming logic without having the additional burden of learning programming language syntax.

In the Scratch interface Blocks palette, blocks are arranged in the following eight categories:

- **Motion**: These blocks enable you to position a sprite on the Stage and optionally move or glide it around. Click Edit, Show Motor Blocks to reveal extra blocks intended for use with the LEGO Education WeDo Robotics Kit (http://is.gd/HIquiE).

- **Looks**: These purple blocks allow you to change the look (called a Costume) of your sprites. You also can have a sprite "say" or "think," as well as ask the user for feedback.

- **Sound**: These pink blocks give you control over system volume and allow your sprites to make sounds of their own.

- **Pen**: These dark green blocks enable your sprites to draw vector lines onscreen.

- **Control**: These gold blocks represent the brains of your Scratch app. You can start scripts, stop scripts, and manage all events within the program by using Control blocks.

- **Sensing**: These light blue blocks are used to detect input from the user. For instance, you can detect mouse clicks, typed responses, and analog events inbound from a PicoBoard. (I'll tell you more about the PicoBoard momentarily.)

- **Operators**: These light green blocks perform mathematical equations and are also used to handle string data.

- **Variables**: These blocks are used to make two types of variables: traditional variables and lists (formally called arrays).

I don't notice this because I am profoundly colorblind, but I'm sure you observed that blocks within each type are color-coded. This helps you associate certain types of actions with certain types of blocks. Moreover, the color-coding helps you keep your variables distinct from each other within your program.

NOTE: WHAT IS A VARIABLE?

A variable is nothing but an in-memory placeholder for a piece of data. Computer programs use variables, which can dynamically change their stored values (hence the name variable), to move data around inside an application.

Scratch blocks themselves fall into six shape types:

- Hat blocks
- Stack blocks
- Boolean blocks
- Reporter blocks
- C blocks
- Cap blocks

Figure 8.5 displays representative examples of each block type. Take a look at them, and then let's learn a little bit more about each block shape.

FIGURE 8.5   A mash-up showing you the different types of block shapes in Scratch. Here's a key to the annotations: 1: Hat block; 2: Stack block; 3: Boolean block; 4: Reporter block; 5: C block; 6: Cap block.

Hat Blocks

Hat blocks have rounded tops, which indicate that they are used to initiate actions, not follow other actions. The Green Flag hat block is universally used to start scripts. I myself also use the Broadcast hat blocks a lot to send and receive messages among different parts of my Scratch app.

Stack Blocks

Stack blocks typically form the bulk of your Scratch programming logic. You can see by the notch at the top and bump at bottom (like interlocking puzzle pieces) that Stack blocks can have blocks attached above and below, forming, well, stacks of programming logic.

Boolean Blocks

Boolean blocks are used to represent binary (yes/no, on/off, true/false, 0/1) conditions in your program. You'll note two things about Boolean blocks:

- They have sharp edges.
- They cannot be stacked, but instead are placed inside of Stack blocks as arguments.

Because Boolean blocks report values (namely true or false), they are also considered Reporter blocks.

Reporter Blocks

Reporter blocks hold values. Like Boolean blocks, Reporter blocks fit inside of other blocks rather than stack themselves. Visually, Reporter blocks have rounded ends as opposed to the sharp ends of Boolean blocks.

C Blocks

C blocks derive their name from their visual appearance. These blocks wrap around one or more other blocks. For instance, you can use a Forever C block to perpetually repeat one or more actions throughout the runtime of the application.

Alternatively, you can apply true/false conditions to C blocks such that their enclosing actions run only as long as the root expression evaluates to True.

Cap Blocks

Cap blocks are used to stop individual scripts or all scripts within the app. You'll see visually that Cap blocks have smooth bottoms and notched tops, which clues you in instantly as to their purpose.

Crafting a (Very) Simple Scratch Application

You will create a fleshed-out Scratch application in Chapter 9, "Programming Raspberry Pi with Scratch—Next Steps." In the meantime, I would be remiss as your guide if I didn't give you some preliminary direction on how to actually build a Scratch game.

Let's get this party started, shall we?

TASK: CREATING A BASIC SCRATCH APP

As you'll be able to see from the following steps, the Scratch project development workflow is friendly, intuitive, and fun:

1. Open Scratch and start a new file. Make sure to save your work—there is nothing more annoying than unnecessarily losing your progress because you forget to save.

2. Let's change the background to something more educational. Double-click the Stage sprite, navigate to the Backgrounds tab, and click Import.

3. In the Import Background dialog box, select the xy-grid background and click OK. Next are some relevant points I want to draw your attention to before we proceed any further:

- The Scratch Cat is the default sprite for any new Scratch app. You can delete it or any other sprite by right-clicking it from the Stage or the Sprites area and selecting Delete from the shortcut menu

- Although you can view the Stage in three different sizes, the Stage itself is of fixed dimensions: 360 pixels tall by 480 pixels wide. In point of fact, the reason I had you load up the xy-grid background is to see these dimensions by using the Cartesian x/y coordinate system.

- What we are going to have this app do is glide the Scratch Cat counter-clockwise, draw a square, and then notify the user that the program is finished.

NOTE: ABOUT THE CARTESIAN COORDINATE SYSTEM

The Cartesian coordinate system represents a handy way to represent two-dimensional space. The X-axis represents the horizontal plane, and the Y-axis represents the vertical plane. You can read a nice write-up on the Cartesian coordinate system at Wikipedia: http://is.gd/93qUQC. In the meantime, know that the notation (100, -100) represents x=100 and y = -100. The built-in xy-grid Stage background in Scratch makes this easier to visualize.

4. Double-click the Scratch Cat sprite, navigate to the Scripts tab, and assemble the blocks as shown in Figure 8.6.

NOTE

When you drag a block in proximity of another block, you'll see a horizontal line letting you know where the block will be positioned. Simply use drag-and-drop to accurately place or move the blocks.

To unlink blocks, note that you need to click and drag the block beneath the block from which you want to detach it. Yes, the block drag-and-drop thing requires some patience—stick with it!

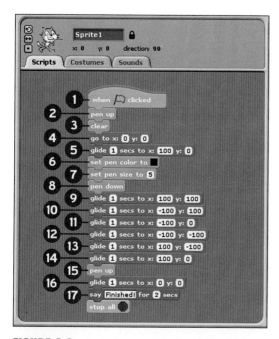

FIGURE 8.6 *Your very first Scratch application.*

Let's explain the purpose of each block in this first app; use the annotations in Figure 8.6 as your reference:

1: Starts the script when the user clicks the Green Flag.

2: Lifts the Pen tool from the Stage (this ensures that the Pen doesn't continue to draw from previous runs of the program).

3: Clears any Pen tool remnants from previous runs of the app.

4: Positions the Scratch Cat sprite instantly at the origin point (x=0, y=0) on Stage.

5: Moves the sprite slowly from origin (0,0) to Stage coordinates (100,0). For more information on the coordinate system see the sidebar, "About the Cartesian Coordinate System."

6–8: Here you customize the Pen tool color and line size. The pen down action figuratively puts down the Pen tool "point" to the Stage surface. Note that you need to manually "lift" the Pen tool with the pen up block to stop the line drawing.

9–14: Here you glide the sprite counter-clockwise around the Stage in a square shape. You should see the line automatically reset and redraw each time you click the Green Flag button to rerun the app.

15: Lifts up the Pen tool and therefore stop drawing lines.

16: Resets the sprite's position to the (0,0) origin point.

17: Has the sprite communicate the end of the program to the user. In this case, the game simply stops. As you gain expertise with Scratch, you will doubtless implement more elegant methods to start, run, and stop your projects.

Then the final block stops the application.

The PicoBoard

Traditional Scratch programming involves sensing and responding to a variety of events:

- Mouse clicks
- Individual keystrokes
- Keyboard-based user input

However, each of these events is what we can call *digital*. In other words, a mouse has either been clicked or it has not—there are no in-between states. Likewise, a mathematical calculation results in a particular result—there isn't any gray area to speak of.

The PicoBoard is a separate piece of hardware you can use to bring analog, external events to your Scratch projects. Fans of the PicoBoard stress that the board allows you to connect your Scratch projects to the outside world.

For instance, how about a game that responds to voice input? Or perhaps a game that uses a custom joystick controller? You can do all this and more with the Picoboard.

The PicoBoard is, like the Raspberry Pi, a printed circuit board. As you can see in Figure 8.7, the PCB consists of a number of inputs that cover a wide variety of analog sensory data:

FIGURE 8.7   The PicoBoard is a sensor module that brings the external environment to Scratch.

1: 4 Expansion Connectors: Each plug links to a pair of alligator clip connectors that can be used to measure resistance in any external object. Scratch represents the connector states as 100 (no circuit) to 0 (complete circuit between alligator clips). Intermediate values represent the degree of resistance in the circuit between the two alligator clips.

2: Slider: Scratch quantifies the slider position in the range 0–100.

3: Light Sensor: The sensor is quantified (or rendered digitally) by Scratch in a range from 0 (totally dark) to 100 (maximum brightness detected).

4: Microphone: This sound sensor is quantified in Scratch in a range from 0 (silence) to 100 (loudest audio signal detected).

5: Button: The tactile (physical) button has two states: True (when pressed) and False (when unpressed).

6: USB: This port both provides power to the board and serves as a way to transfer data to and from the PicoBoard.

Incidentally, analog signals are distinct from digital ones because analog signals operate on a continuum of continuously varying values. For instance, the human voice generates a wave-like pattern of data. When computers use analog-to-digital converters (which the Pi can do thanks to the Gertboard accessory), they attempt to replicate an analog waveform by using two values: 0 and 1. The more bits you add to the conversion, the more faithfully you can reproduce the original audio signal. That's why low bitrate MP3 audio sounds so much worse than high bitrate MP3 audio.

Obtaining a PicoBoard?

The PicoBoard is manufactured and sold by Sparkfun Electronics (http://is.gd/WyVO4D); as of this writing, the price is $44.95. Please note that the PicoBoard as it is sold by Sparkfun does *not* include the requisite mini USB cable that is required to use the board.

Note, also, that the PicoBoard uses the mini USB cable, not the micro USB cable that the Raspberry Pi uses.

Using a PicoBoard in Scratch

Recall that the Raspberry Pi requires at least 700 mA inbound to perform its work and that you should plan on using a powered USB hub to power any external devices.

To that point, be sure to plug your PicoBoard's USB cable into your powered hub and not into the Pi itself.

I've found that the Raspbian OS automatically detects the PicoBoard, and the device is therefore immediately usable in the Raspberry Pi. This is good news because you need to manually install PicoBoard device drivers for Windows and OS X computers (you can download the drivers from the Cricket website at http://is.gd/GTkHm7).

TASK: USING THE PICOBOARD IN SCRATCH

Now that I've whet your appetite and you've received your PicoBoard, it is time to learn how to use it with your Raspberry Pi. Let's begin!

1. Ensure that your PicoBoard is plugged in and that Scratch detects it. An easy way to test functionality is by using the ScratchBoard watcher.

2. Navigate to the Sensing blocks palette, right-click the slider sensor value block, and select show ScratchBoard watcher from the shortcut menu.

   This action adds a Stage monitor (also called a watcher) to the Stage that displays the current values of all sensors on the PicoBoard. You can see this in action in Figure 8.8.

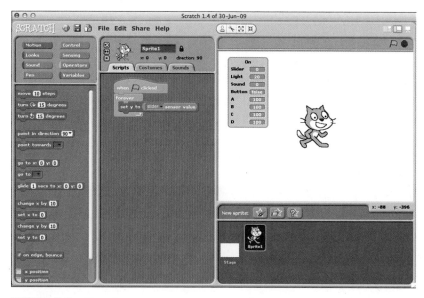

FIGURE 8.8   The ScratchBoard Watcher gives you at-a-glance status of all PicoBoard sensors. You can see the Watcher in the Stage area above and to the left of the Scratch Cat sprite.

3. In Raspbian, open a web browser and download the Scratch project called "PicoBoard Tester" (http://is.gd/ry5nra).

4. In Scratch, click File, Open and navigate to the PicoBoard Tester project. Open the project and click the Green Flag icon to run it.

5. On the PicoBoard, jog the slider back and forth. Note both the graphical element on the Stage as well as the watcher readout value.

NOTE: THE DIFFERENCE BETWEEN THE SCRATCHBOARD AND THE PICOBOARD?

The MIT Learning Lab people originally developed the PicoBoard; the device was initially yellow and carried the name Scratch Sensor Board. Before too long, ownership of the project changed hands a couple of times. Now Sparkfun Electronics owns the hardware; to celebrate they renamed the device PicoBoard and gave the PCB a nice red paint job.

6. Snap your fingers. Again, observe both the program's graphical display as well as the value on the watcher. You'll find that the PicoBoard microphone is pretty darned sensitive!

7. Pick up your PicoBoard (carefully) and hold it close to a nearby light. By contrast, slowly cover up the light sensor with your cupped hand. Observe value changes in the Scratch project.

8. Click the tactile pushbutton on the PicoBoard and watch for changes in the Scratch app.

9. Finally, plug in one of the alligator clip pairs and touch the alligator clip metal ends together. Note the changes in the Scratch program that indicate you've completed a zero-resistance circuit between the probes.

Two final points to consider regarding this final project:

- Try building your own sprites that take actions based on PicoBoard-detected events.
- Remember to scour the source code of any project you download from the Scratch website. You can learn a lot about best (and worst) practices by studying how other Scratchers think and develop their apps.

## CHAPTER 9

Programming Raspberry Pi with Scratch—Next Steps

If you invested the necessary time to practice the Scratch programming skills you learned in Chapter 8, "Programming Raspberry Pi with Scratch—Beginnings," you are ready to take the next step by developing and sharing a full-fledged Scratch application.

In this chapter, you learn how to create a pretty neat game, if I do say so myself. You've even got my blessing to remix the game and submit it to the Scratch Projects website (if you don't know what remixing means, don't worry—I cover that later on).

NOTE: VERSION CONTROL, REVISITED

As you learned in Chapter 8, the Scratch Team at the MIT Media Lab has upgraded their website, promotional materials, and the development environment itself to Scratch 2.0. Although Scratch 2.0 on its surface looks very different from Scratch 1.4, don't be daunted. Everything you learned in the previous chapter and everything you learn in this chapter carries over root and branch from Scratch 1.4 to Scratch 2.0. Besides, this book is about the Raspi, and Raspi includes Scratch 1.4.

Begin with the End in Mind

Before you begin coding, I think the review of some programming best practices is in order. What questions do you need answers to before you open up your development tools and start to build an application?

What exactly is the purpose of the app? For the purposes of this exercise, you want to build a game in Scratch that tests the player's reflexes and offers a minute or so of heart-racing fun.

Specifically, the game I designed for this chapter is called *Dodgeball Challenge*; Figure 9.1 displays the splash screen for the game.

FIGURE 9.1 Dodgeball Challenge splash screen. By the end of this chapter, you'll know how to build this game!

From a meta perspective, I also intend for this case study Scratch game to serve as a showpiece for what Scratch is capable of as a multimedia development environment.

Who is the intended audience for the app? Speaking personally, I have two main audiences in mind: you, my readers and students, who want to learn how to program in Scratch and anybody with a love of simple, addictive video games.

I have actually added the Dodgeball Challenge game to the Scratch website. If you like, you can visit the game page (http://is.gd/RvvRsT) and check out the game for your reference (see Figure 9.2). Isn't open source software wonderful?

Before you proceed to the rest of the chapter in which you build the game from Scratch (pun intended), play the game a few times so that you're familiar with the gameplay. Although I provide the game rules on the Scratch website page, let me summarize them for you here for the sake of completeness:

- The goal of Dodgeball Challenge is to survive for 60 seconds.
- The game ends if the user's avatar (specifically, the Scratch Cat as controlled by your computer mouse) touches a ball or (optionally) the edge of the playfield.
- Every 15 seconds a new ball is added to the game, ramping up the complexity and difficulty.

FIGURE 9.2 You can check out the Dodgeball Challenge game (and associated source code) from the Scratch Projects website. Note that the game is automatically converted to Scratch 2.0 and can be opened directly from a (non-Raspberry Pi) web browser.

Let's Build the Game!

Fire up Scratch and save a new .sb project file. Because the Scratch Cat is the default sprite, let's go ahead and use it. The first step in this process is to set up the game screens, but before you do that, let's set the table with regard to what you're about to take on.

We have a concrete idea for a fun game, and we'll build it one piece at a time, starting with the screens. *Iterative software development* means that you'll test the game after every change. This approach minimizes the possibility of introducing bugs (programming logic flaws) into the released version of the game.

Take a look at Figure 9.3 so that you can refamiliarize yourself with the Scratch 1.4 user interface.

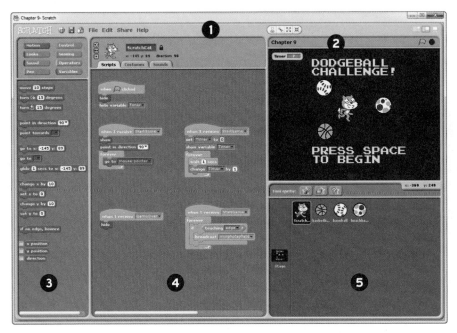

FIGURE 9.3   The Scratch 1.4 user interface, revisited.

1: Menu and toolbars

2: Stage

3: Blocks palette

4: Script area

5: Sprite area

TASK: SETTING UP THE GAME SCREENS

In Scratch 1.4, the Stage is the graphical area in which all activity takes place. Recall from the previous chapter that the Stage consists of one or more Backgrounds that function a lot like PowerPoint slides. Your first development task in building Dodgeball Challenge is to define those game screens. We'll worry about wiring them together afterward.

1. In the Sprites area, double-click the Stage to bring it into focus. In the Scripts area, navigate to the Backgrounds tab. Use the Paint button to create three additional playfield screens (four in total) as shown in Figure 9.4.

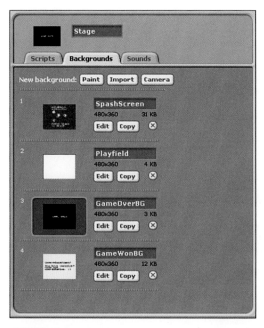

FIGURE 9.4   Dodgeball Challenge Stage backgrounds.

Here's the breakdown of the purpose of each background:

- **SplashScreen**: This is the introductory screen that advertises the game to players and provides instructions on starting the game itself.

- **Playfield**: This is the screen that is used during actual gameplay.

- **GameOverBG**: This is the screen the player sees when he loses the game.

- **GameWonBG**: This is the success screen presented to the player who lasts the entire 60 seconds without losing.

For further assistance in creating your game screens, feel free to study the backgrounds in my published copy of Dodgeball Challenge on the Scratch website at http://is.gd/RvvRsT. Because my published version uses Scratch 2.0, you can view the underlying source code and assets directly from your web browser.

Another cool tip I have to share is that you can freely and legally download the beautiful Press Start 2P 8-bit retro videogame font from the FontSpace website (http://is.gd/59fciQ).

2. Navigate to the Scripts pane and add the two code blocks shown in Figure 9.5. Ignore the two "when I receive" stacks for now.

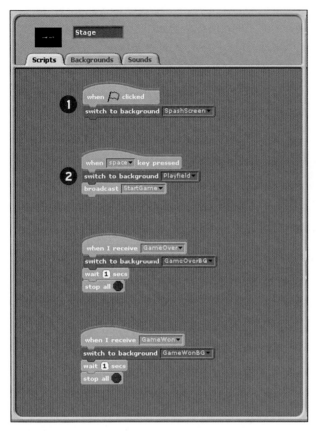

FIGURE 9.5   Code blocks for the game Stage.

1: This block ensures that when the player clicks the Green Flag, the black splash screen is presented.

2: This block triggers the actual gameplay by switching the background to the white playfield background and kicking off the StartGame broadcast. Note that I chose the Spacebar as the method for starting the game, but you can select any key from the keyboard. Although you can also set up the game to respond to a mouse click, I advise against that for this broadcast so as not to introduce any potential confusion. For my part, I tend to use out of the way keystrokes in my games so if the user clicks their mouse or taps a common key (such as ENTER), the game does not blow up.

As I alluded to in the previous chapter, broadcasts provide a convenient method for communicating among sprites (or between the Stage and sprites) in a Scratch application. You can use the broadcast block from the Control palette to define a broadcast; make sure to give each broadcast a meaningful name.

You can then leverage the when I receive block (again from the Control palette) to receive, or catch, outbound broadcasts from the same sprite, another sprite, or even the Stage.

Let's now turn our attention to initial setup of the ScratchCat sprite.

TASK: SETTING UP THE SCRATCH CAT

1. Double-click the Cat to select the sprite and rename it ScratchCat in the Scripts area.

2. Click the Shrink Sprite button above the Stage and repeatedly click the Scratch Cat sprite until it reaches your desired size. My thought was to make the sprite large enough to see its detail, yet small enough to provide for challenging gameplay.

3. Go over the Scripts area and set up the block stacks as shown in Figure 9.6. Yes, I know you can simply examine my source code from the game copy you downloaded from the Scratch website. However, if you want to learn to program with Scratch, you need to test this stuff out yourself!

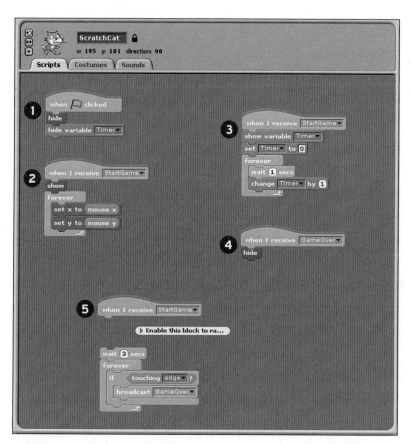

FIGURE 9.6 Block setup for the ScratchCat sprite and the game timer.

Let me explain the purpose of each block stack:

1: This stack hides both the Scratch Cat as well as the timer that you are about to build when the player clicks the Green Flag. You don't need to see these elements until the player actually starts the gameplay itself (that is to say, when the player taps the Spacebar).

2: This event listener "unhides" the Scratch Cat and maps the player's mouse movements to the Scratch Cat's location. Note that these actions kick off in response to the StartGame broadcast that you initiated from the SplashScreen.

3: This block stack sets up the game timer, which is crucial in this game. Although the Sensing palette contains a couple Timer blocks, those blocks mark the elapsed time since you last opened the Scratch app. Of course, that's not what you need here. Therefore, you can create a new variable named Timer (discussed in the Note, "Creating Variables") and simply increment its value every second. Pretty straightforward stuff, wouldn't you agree?

NOTE: CREATING VARIABLES

Variables are how you store temporary data in your app. To create the Timer variable, navigate to the Variables palette and click Create Variable. By default the new variable will be available to all sprites in the game; that is what we want in this case. For further details, look at the live source code at the Scratch website: http://is.gd/RvvRsT.

4: This is an optional block stack I put together to give you the ability to ramp up the game's complexity. This block ends the game if the Scratch Cat touches the border of the screen. I disabled the stack by default because I felt that it made the game overly difficult. What do you think?

5: When the GameOver message is broadcast, I want the Scratch Cat sprite to disappear.

Setting Up the First Ball

I chose to add most of the gameplay logic in the Scripts area for the Basketball sprite, which is the first enemy sprite used in the *Dodgeball Challenge* game. Take a look at the code in Figure 9.7, and I'll walk you through it.

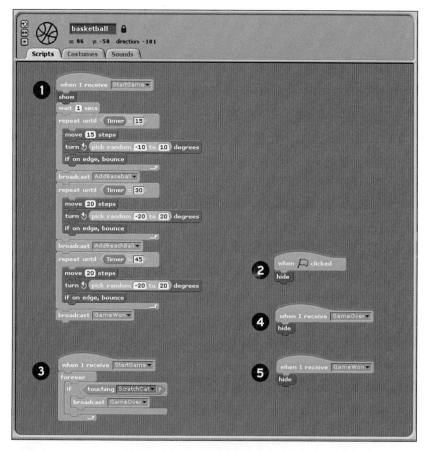

FIGURE 9.7   Block setup for the Basketball sprite.

1: Wow, this is one honkin' block stack! Here's the deal: When the user presses the Spacebar and kicks off the StartGame broadcast, you want to perform the following actions:

- Show the previously hidden basketball sprite (I prefer doing this to having the ball materialize out of thin air).

- Wait an arbitrary second to give the player their bearings before the game starts.

- Loop through "random" movement. If you look within each repeat block, you see that I have the ball move 15 steps, turn in a random direction between two degree markers, and bounce if the sprite hits the border of the Stage.

- When the timer reaches 15, 30, and 45 seconds, I introduce additional balls. These invocations are handled, reasonably enough, through broadcasts.

- When you reach 60 seconds without a game-ending action, Scratch broadcasts the GameWon message and congratulates the player on successfully completing the challenge.

- You'll also note that I cranked up the speed and behavior of the basketball at the 15, 30, and 45 second marks by adjusting the step count as well as its directionality.

3: This code stack says that if the ScratchCat sprite makes contact with the basketball sprite, then the game ends by broadcasting the GameOver message.

2, 4, 5: These code blocks state that the basketball should be hidden when the player clicks the Green Flag or when the game ends, either successfully or unsuccessfully. The main point here is to let you know that a sprite can listen to and respond to its own broadcast messages.

Setting Up the Second and Third Balls

The setup for the Baseball and Beachball sprites is much easier than that of the basketball because we've already handled their introduction to the game. Look at the code in Figure 9.8 and see for yourself:

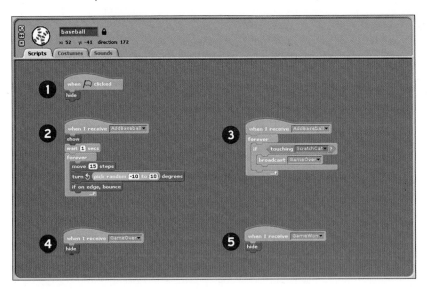

FIGURE 9.8   Block setup for the Baseball sprite; the Beachball sprite is set up the same way.

1: You want the ball to hide when the player initially runs the game.

2: When it's time to add the second and third balls, this block performs the same show/ wait 1 second/start to move actions. The difference here is that you aren't linking any additional behavior to the second and third balls (but go ahead if so inclined; I just thought the game was difficult enough as it was).

3: I discussed this block already; you want the game to end if the ball and the ScratchCat touch.

4, 5: These blocks instruct the balls to disappear when the game ends.

Debugging and Troubleshooting

With almost any computer program, you need to be on the lookout for so-called bugs. Bugs can take the form of logic problems, syntax errors, missing references—unfortunately, the list of possible bug sources is almost limitless.

The good news is that software development industry has established best practice for identifying, trapping, and resolving software bugs. Even Scratch includes built-in debugging functionality.

To these points, following are some good best practices for writing Scratch games that are as close to "bug free" as possible.

- **Run, re-run, and re-re-run your project.** This is called *iterative application development*, and it is crucial for you to do to ward off any glitches that will anger and frustrate your users. The Scratch environment makes it really easy to start, stop, and restart your project. Thus, you should get into the habit of testing your changes as you make them.

- **You can run individual code stacks.** Instead of using the Green Flag to run your project from start to finish, you can test the behavior of individual code stacks simply by clicking its Hat block. Try it out—it's helpful from a troubleshooting and debugging standpoint.

NOTE: THE ORIGIN OF THE SOFTWARE BUG

Why are problems with software applications (errors, flaws, or failures that cause unexpected results) called "bugs"? The term was first used in the context of hardware engineering by the inventor Thomas Edison, who wrote in an 1878 letter, "It has been just so in all of my inventions. The first step is an intuition, and comes with a burst, then difficulties arise—this thing gives out and [it is] then that 'Bugs'—as such little faults and difficulties are called—show themselves and months of intense watching, study and labor are requisite before commercial success or failure is certainly reached."

- **You can enable Single Stepping mode.** In Scratch, click Edit, Set Single Stepping. You'll be presented with a list of speeds, from Turbo on one extreme to Flash blocks (slow) on the other.

    Single Stepping gives you the ability to slow down the execution of your Scratch game so you can better see code stacks and individual blocks firing. This makes it easier to see if, when, where, and why your program does not behave the way you want it to.

After you choose your Single Stepping mode, click Edit > Start Single Stepping to begin your debugging session. Be sure to click Edit > Stop Single Stepping when you are finished to return Scratch to its "factory default" behavior.

- **Get block-level help.** Try right-clicking a block that's in your project and selecting Help from the shortcut menu. You will see a nifty Help window that graphically shows you how the selected block functions in the context of a practical example (see Figure 9.9). I personally find this flavor of online help to be invaluable because sometimes I discover the perfect block that I previously overlooked, which solves my current troubleshooting or debugging problem.

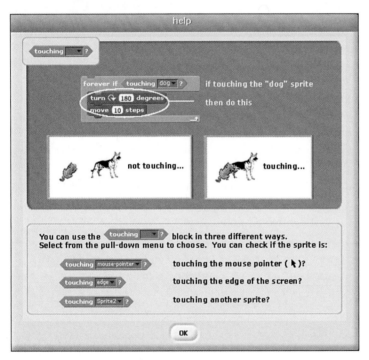

FIGURE 9.9 Scratch is a beautifully documented coding environment.

- **Consider documenting your code.** Right-click near a block in your Scripts area and select Add Comment from the shortcut menu. Documentation is one of the most important best practices in all of computer programming.

    These comments can be intended simply to remind you to fix something a little bit down the road. You can also use comments to explain your rationale in doing something in your program (see Figure 9.10); this can be helpful in a year or so when you re-open the project and think to yourself, "What in the world was I thinking when I wrote that?!"

FIGURE 9.10 Commenting your code is a programming best practice.

Finally, adding comments to your code helps make your code more understandable to fellow Scratchers who download and inspect your project. Note that you can link a comment to a block by dragging the comment in the proximity of said block. The little disclosure arrow enables you to shrink or expand any comment.

Uploading Your Project

Now that you've completed your project, it is time to share your work with the Scratch community. Why do this? Here are a few good reasons:

- **You're proud of your work.** You just invested time and effort in creating your Scratch game. Thus, you want to get the application in front of as many sets of eyeballs as possible. This is a natural and honorable motive of any self-respecting software developer.

- **Share and share alike.** The spirit of open source software is to share your work with other developers to solicit their feedback. By contrast, you are expected to offer constructive criticism of other peoples' Scratch projects. That's how we all learn!

TASK: SHARE YOUR SCRATCH PROJECT

Sharing your Scratch project with other Scratch users around the world is the best way to get your game (and underlying code) in front of as many eyes as possible. You'll be able to garner feedback from other Scratchers, and the sense of accomplishment you'll feel when other users create remixes of your work is indescribable.

1. With your project open in Scratch, click Share, Share This Project Online or click the Share This Project button on the main toolbar. The interface is shown in Figure 9.11.

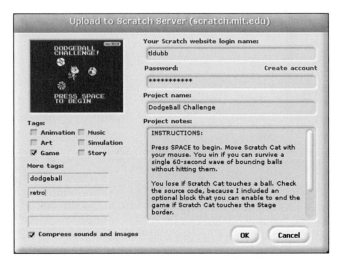

FIGURE 9.11 You can upload your Scratch project to the Projects website directly from within the Scratch application.

2. Fill in the fields in the Upload to Scratch Server (scratch.mit.edu) dialog box as completely as possible. Let me help you understand what's involved:

- **Your Scratch website login name and Password**: As with most things Scratch-related, you must have a free Scratch account to upload a project. Click Create account to, well, create a new user account.

- **Project name and Project notes**: These fields are automatically populated based on the .sb project file name and any project notes you added. I highly recommend you add project notes (welcome message, instructions, and so on) to help people understand and use your Scratch app more efficiently and effectively. You can add your project notes either in this dialog box or by clicking File, Project Notes in Scratch.

- **Tags; More tags**: Tags are keywords that make it easier for Scratchers to find projects on the Scratch website. You can select any of the six prebuilt tags and/or define your own.

- **Compress sounds and images**: This option is a good idea because it makes your Scratch program smaller. A smaller .sb file means that the app runs faster in the web browser and takes less time for your users to download from the Scratch website.

As you saw in Figure 9.2, the MIT Media Lab developers give you a dedicated web page for each app you upload. Be sure to check out the page and perform the following actions on a regular basis:

- Read comments left by other Scratchers and take their criticism constructively.

- Proofread and potentially edit your project notes. You can do this directly from within the web browser, provided you are logged into the site with your Scratch account.

- Consider adding additional tags as they occur to you. There is no upper limit on how many tags a Scratch project can have associated with it. To that point, other Scratchers can tag your project as well.

Remixing

Sometimes you'll come across a Scratch project that is so good it leads you to think, "I'll bet I can make this good app great!" This notion is perfectly legal as well as in keeping with the open source community spirit shared by Scratch and other public domain frameworks.

In Scratch nomenclature, a *remix* is a Scratch project that is based on somebody else's Scratch project. When you publish a remix, a link to your remix is accessible on the original app's home page. By contrast, viewers of your app can click a link that takes them back to the original app's home page at scratch.mit.edu.

Remember that the name of this program, Scratch, derives from the disc jockey (DJ) term of moving vinyl records back and forth to create rhythmic sounds. Similarly the term remix refers to the method by which musical artists re-record their songs by using the same melody but changing the style.

NOTE: ABOUT LICENSES...

Scratch apps fall under not the traditional GPL license like Raspberry Pi does, but instead under the related Creative Commons Attribution-ShareAlike 2.0 license. You can read the user-friendly license details at the Creative Commons website at http://is.gd/FAsiS7).

TASK: CREATE A REMIX

1. Download another Scratcher's project from the Scratch Projects website.

2. Modify the project as needed.

3. Upload the project using the method described earlier in the task, "Share Your Scratch Project." The bi-directional linking and notifications between the original author's project and your remixed project happens automatically, or as my old friend Jeff Kane used to say, "automagically."

## CHAPTER 10

Programming Raspberry Pi with Python—Beginnings

If you emerged from Chapters 7 and 8 relatively unscathed, you are ready to embark on a more comprehensive programming adventure.

Scratch is a lot of fun to program, but the environment hides all of the programming complexity; this is by design, as previously discussed. Many educators consider Python to be an ideal first "true" programming language for the following reasons:

- Python's syntax and data typing are relatively intuitive and fairly forgiving of rookie mistakes.
- Python is heavily documented; you can find easy-to-follow tutorials just about anywhere.
- Python's interactive interpreter makes learning new stuff fast and fun.
- Python offers an amazing number of importable code libraries that give beginning programmers tools to build any kind of application.

Python is called a "general-purpose, high-level programming language" whose overarching design principle is code readability. In fact, you would be well-advised to read what Python fans consider to be their fundamental, guiding principles: the Zen of Python (http://is.gd/sXV6IU). Let me share with you my favorite entries from the Zen document:

- Explicit is better than implicit.
- Simple is better than complex.
- Readability counts.

For three simple sentences, that's quite a bit of wisdom, right? As a programmer, you are much better off writing code that is as straightforward as possible and documented in such a way that any other Python programmer can read your code and instantly understand how your program works. To be sure, if you've been practicing with Scratch, you already understand how important community support is when you're developing software projects.

Python's focus on clarity and readability probably weighed heavily into the Raspberry Pi Foundation's decision to build the Raspberry Pi development platform around Python. If you ever saw C or C++ code, you will instantly appreciate how much more approachable Python is on almost every level.

NOTE: WHERE IT BEGAN

Just a tad bit of history before we dive in: The Python programming language was invented by the Dutch programmer Guido van Rossum in the late 1980s. Rossum needed a fast, intuitive scripting language to help him automate administrative tasks, and he wasn't getting very far with the tools he had in front of him at the time. Thus, Guido adapted the ABC programming language that was popular in the Netherlands at that time to a new language that focused on simplicity and readability without sacrificing power—enter Python!

What's so cool about Python is its flexibility—some call Python a scripting language because you can write and test code quickly without the need for binary compilation. On the other hand, because Python has grown into a robust language that supports enterprise-level concepts such as object orientation, the term high-level programming language seems more appropriate for Python.

The way I want to teach you Python in this chapter and the next one (itself a hugely daunting task), is to jump right in and get your hands dirty with the environment. At the end of this chapter I give you some hand-selected online and print resources with which you can take the next steps in your development as a Python programmer.

To that point, however, I strongly encourage you to pick up *Sams Teach Yourself Python Programming for Raspberry Pi,* written by my Pearson colleagues Richard Blum and Christine Bresnahan. Their book touches briefly on the material we deep-dive into (the innards of the Pi), while my book does the same thing with regard to Python programming. I think that Richard and Christine's book and my book complement each other quite nicely, thank you very much!

Let's get to work.

Checking Out the Python Environment

Boot up your Pi and fire up a Terminal prompt. It doesn't matter whether you are in LXDE or not at this point, although you'll need LXDE in time when we discuss IDLE.

As it happens, there are two versions of Python currently in use today, and both of them are included in the Raspbian Linux distribution. In this book I choose to stick with Python 3, the current version, because it is a nice improvement over Python 2 (for those interested in an exhaustive comparison, visit the Python website at http://is.gd/kYsc97).

Try out the following commands, pressing Enter in between:

```
python -V
python3 -V
```

What you just did was to check the currently installed versions of Python 2 (the executable program file is python) and Python 3 (executable program file name of python3). As with all things in Linux, case is sensitive, so you need to supply the -V parameter and not -v to see the Python version.

Later in the book, you'll find a lot of the code you need to undertake certain Raspberry Pi projects was written in Python 2. Don't be alarmed! For our purposes, the differences between Python 3 and Python 2 are under the hood and everything you learn in this chapter and the next one is directly applicable to the Python 2 work you'll do later on.

Now pop into LXDE and look at the two icons labeled IDLE and IDLE 3. These are shortcuts that open the **I**ntegrated **D**eve**L**opment **E**nvironment, or IDLE (pronounced *eye-duhl*), for Python 2 and Python 3, respectively. Try double-clicking IDLE 3; the interface can be found in Figure 10.1.

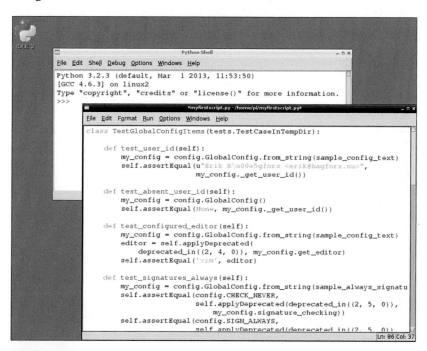

FIGURE 10.1   The IDLE development environment for Python 3.

What's cool about IDLE, also called the Python Shell, is that it is itself a Python application that leveraged the Tkinter (pronounced *tee kay inter*) GUI toolkit. Tkinter and packages like it enable you to build graphical Python applications that leverage windows, colors, icons, buttons, and mouse navigation. That said, we're focusing on console (command-line) applications in this book because Python graphical programming is an advanced topic and warrants its own chapter if not its own book.

An Integrated Development Environment, or IDE, is useful to a programmer because most IDEs offer time-shaving functionality such as the following:

- Interactive help with programming language syntax
- The ability to step into programs and stop/restart at predefined points
- Detailed insight into design-time and run-time errors

IDLE offers all of this and more. It's definitely not the most robust (or even stable) IDE, but I use it here because it comes standard with Python, and it's already loaded in Raspbian.

NOTE: PYTHON IDE ALTERNATIVES

If you discover that you love Python and want to try out alternatives to IDLE, be sure to check out some of the open source and commercial code editors and full-fledged IDEs that are out there. Some notable examples include Eclipse IDE (http://is.gd/fGo9FH) with the PyDev extension (http://is.gd/yDXZ8m), Komodo IDE (http://is.gd/blPFT9), and WingIDE (http://is.gd/jyg8ig).

Enough about IDLE. The whole of Chapter 11, "Programming Raspberry Pi with Python—Next Steps," is dedicated to building Python programs using IDLE. For the remainder of this chapter, we'll stay in the Terminal environment to interact with Python 3.

Interacting with the Python 3 Interpreter

Open up a Terminal session on your Pi and try out the following procedure. Be sure to press Enter after issuing each command.

```
python3
print("Hello world!")
```

In the first line of code, you invoked the Python 3 interpreter. This means that until you either close the Terminal window or issue the exit() command, everything you type is sent directly to Python on your Pi.

In other words, when you send a Python statement to the interpreter, Python parses, or interprets, that line of code, performs the instruction(s) contained in the code, and presents the results as appropriate on the screen.

Thus, Python is called an interpreted programming language because it takes your plaintext code input and processes it directly into machine language. (Technically, Python busts the plaintext code into an intermediate state called byte code, but we don't need to get too picky about it at this point.)

Other popular interpreted programming languages include Java, JavaScript, PHP, and Ruby. However, programming with these languages on the Raspberry Pi is likely to be an exercise in futility because (a) you have to install quite a bit of extra software to get those environments functional; and (b) as you know by now, the Raspberry Pi is not exactly a processing workhorse. Java in particular is known for the impact it can have on resource-constrained computer systems. Therefore, the Foundation's decision to standardize on Python is very wise because Python is a low-overhead, flexible, and powerful programming environment.

By contrast, compiled programming languages such as C and Microsoft .NET must be converted into executable binary machine language prior to their being run. Therefore, developing compiled-language projects takes much longer than creating interpreted-language ones because the compilation process can sometimes be tedious and time-intensive.

As I said earlier in this chapter, interpreted programming languages are oftentimes called scripting languages because of their agility and speed at which they can go from design time to run time.

Exiting the Interpreter

You can tell at a glance that you are inside the Python interactive interpreter because the command prompt displays as three right angle brackets (>>>). If you need to exit the interpreter to return to the command prompt, issue the following command:

```
exit()
```

To return to the interpreter, you know the drill: Type **python3** and press Enter.

Getting Online Help

The help system that is built into the Python interpreter is, well, passable, although in my opinion you can't beat the Python Documentation website (http://is.gd/SGejBj; keep it bookmarked!). Nevertheless, to access the online help, issue the following command from within the Python interactive interpreter:

```
help()
```

To get a list of help topics, type **topics**. If you want a list of Python keywords, type **keywords** (see Figure 10.2).

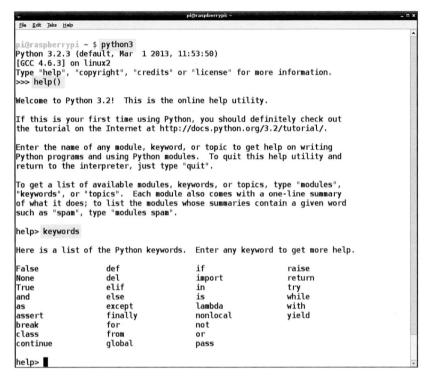

FIGURE 10.2 The Python interpreter online help is decent enough.

You can get help regarding a particular keyword by typing that keyword (try **print**, for instance) from within the online help system. Alternatively, you can jump directly to a specific help page by issuing help('print') from outside the help system but inside the Python interpreter.

Either way, after you are inside a help page and are ready to exit, type **q** (no, this is not intuitive). To add insult to injury, you must remember to press Ctrl+D to exit the online help system and return to the interpreter. Now who said that Python was intuitive again?

Writing Your First Python Program

The traditional first example when a student learns a new programming language is to have the program print "Hello World" on the screen. The remainder of this chapter is in keeping with that tradition.

From the Python interactive interpreter, issue the following command:

```
print("Hello,World!")
```

Congratulations on creating your first Python program! What you did in a single line of code is to instruct Python to output the string "Hello,World!" to the screen. Specifically, print is what's called a function—a *function* is a named object that performs some action when the function is invoked.

Functions typically take one or more input parameters; these are fed to the function inside of parenthesis. Hence, in our Hello World example, Python fed our "Hello, World!" string as an input parameter to the print function, which in turn was written to echo its parameter to whatever output you specify (the screen, also called *standard output* or *stdout*, is used in the absence of a specific output value).

Issuing Python statements from the interactive interpreter is all well and good, but it is not sustainable for anything outside of the smallest of tasks and for testing/diagnostic purposes. To preserve your hard development work, you need to save your Python source code in a script file.

Historically, Python script files use the file extension .py. Under the hood, however, these are plaintext files that are creatable and readable in any text editor. Today we'll use (you guessed it)...nano.

TASK: CREATING YOUR FIRST PYTHON SCRIPT

1. From a Terminal session, ensure that your present working directory is your home directory:

```
cd ~
```

2. Create a new, blank text file in your home directory using nano as your editor:

```
nano firstscript.py
```

3. Add the following code (which you can also see in Figure 10.3):

```
1: #!/usr/bin/env python
2: fn = input("What is your first name? ")
3: print("Hi there," , fn, "\n")
```

FIGURE 10.3 Your first Python script.

Let's cover what each line in the script means:

- **1**: This is called the *shebang line*, and it simply instructs Linux as to where to find the Python executable program file. This is useful so you can invoke the Python interpreter from whatever present working directory you may be in at a given time.

- **2**: This creates a variable named fn that stores the user's response to the question string *"What is your first name?"* A variable is simply a temporary, in-memory placeholder for data. Because Python is a dynamically typed language, you don't have to specify the data type for our new variable; the interpreter sees that you are obtaining string data and formats the variable accordingly.

    The input function is used to solicit feedback from the user. The input parameter is simply the prompt string. Notice that I added an extra space after the question mark and before the closing quotes—this is to make the question and the user's response more legible.

- **3**: This command invokes the print function to present a dynamic string to the user. Use the comma inside the print function arguments to *concatenate*, or combine, multiple strings. Note that in this example the line concatenates three discrete elements:

    - The static string "Hi there,"

    - The current value of the fn variable

    - A newline character

    Escape sequences are used in Python to send internal commands to the Python interpreter. The \n escape character (all escape characters begin with the backslash, by the way) tells Python to insert a new line at that point.

- **4.** In nano, press Ctrl+X, Y, and Enter to save your work and exit the editor. Now it's time to run the new script.

NOTE: THE MOST COMMON NANO KEYBOARD SHORTCUTS

As you gain more experience with Linux and its myriad text editors, you may (like myself) choose nano as your preferred tool. To that end, you should memorize the following keyboard shortcuts: Ctrl+O (the letter, not zero) to save. Ctrl+V to jump to the next page. Ctrl+Y to return to the previous page. Ctrl+W to perform a keyword search. Finally, there is Ctrl+X to exit.

Running Python Scripts

In Raspbian, the path to the Python interpreter is included in the OS program search path. Therefore, you can invoke Python 3 from wherever you are in the command-line environment. However, you do need to be aware of where your target script is located.

TASK: RUNNING PYTHON SCRIPTS (COMMAND LINE)

When we're experimenting with the Raspberry Pi in projects that are presented later in the book, you'll be running several scripts. Therefore, learning how to execute Python scripts from the command line is a useful skill for any tech enthusiast, much less a student of the Raspberry Pi. Let's begin!

1. Issue **pwd** to check your present working directory. If you aren't in your home folder, issue **cd ~**.

2. To run a .py Python script that exists in the current directory (like your new script should be), run the following command:

```
python3 firstscript.py
```

3. How did your program run—pretty well? Good. Let's now change to a different directory:

```
cd /tmp
```

4. Try running the command in step 2 again. Were you successful? No? Now try the following:

```
python3 /home/pi/firstscript.py
```

Cool. At this point you should have a pretty good feel for how to run Python scripts from the command line.

TASK: RUNNING PYTHON SCRIPTS (IDLE)

I know that I said earlier that we focus on the IDLE environment in the next chapter, however, as long as I'm already on the subject it makes sense that I cover running scripts in IDLE now.

1. From the Raspbian LXDE desktop, double-click IDLE3 to open the Python 3 editor.

NOTE: VERSION CONTROL, RE-REVISITED

Remember that I'm using Python 3 in this book, so make sure you open IDLE 3 and not IDLE. You'll immediately know you've invoked the incorrect Python version because you'll see errors related to the print function, which underwent a change from a simple statement to a formal function between Python 2 and Python 3.

2. In the Python Shell window, click File, Open.

3. Use the controls in the Open dialog box to select the target .py script file. I show you this interface, which should be immediately understandable to OS X and Windows users, in Figure 10.4.

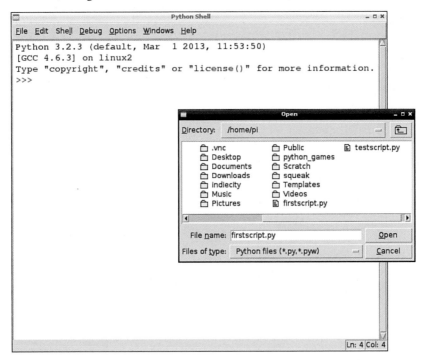

FIGURE 10.4 IDLE behaves like most GUI editor programs.

4. You'll see your script show up in a separate IDLE editor window. To actually run the script, simply click Run, Run Module or press F5.

Broadening Your Python Horizons

Many programmers, myself included, learn a great deal concerning best and worst practices by studying the code of other developers. To that end, I want to share with you some rich sources of example Python scripts you should download to your Pi, run, and analyze.

For instance, here are links to some excellent Python sample code repositories that ought to give you plenty of experience and fun:

- **Code Like a Pythonista**: http://is.gd/fE2Owx
- **LearnPython.org**: http://is.gd/BtqwhA
- **Dive into Python 3**: http://is.gd/3tP9ZL

I want to give a shout-out to Professor Anne Dawson of Coquitlam College in Canada: She put together an excellent list of Python 3 example programs at http://is.gd/Ipv80w. You'll note that the file is a plaintext text file, which means you can easily copy and paste her code snippets into your own environment without carrying any extra HTML formatting baggage.

There are, however, a number of Python community websites that are static, and every budding Python programmer should have them bookmarked and review them frequently. Here are my own hand-picked selections:

- **http://is.gd/Pf9vb4**: CPython is the standard Python distribution.
- **http://is.gd/SGejBj**: Python 3 official documentation.
- **http://is.gd/EWR7d3**: Python Enhancement Proposal (PEP) Index—PEPs are documents that define the formal Python specifications and best practices.
- **http://is.gd/sXV6IU**: PEP 20 is called "The Zen of Python" and is required reading for any aspiring Python programmer.
- **http://is.gd/nCexcw**: PEP 8 is titled "Style Guide for Python Code." You'll find this reference to be increasingly useful as you gain experience with Python programming.
- **http://is.gd/nNYCUy**: *Learning Python* by Rick Lutz is, in my humble opinion, the best Python beginner's book on the market.

So what do you think of Python as compared to Scratch? Are you able to see how Scratch projects are directly analogous to Python programs, albeit with greater simplicity?

At base, all computer programs behave the same way as computers: they accept instructions as input, perform some processing on that data, and then produce output to the user.

Moreover, all computer languages, no matter how rudimentary or cryptic their syntax rules, do the same kinds of stuff: the concepts of variables, procedures, loops, debugging, interpretation, compiling, and execution are the same no matter what specific language you feel most comfortable with.

In the next chapter, I help you broaden and deepen your understanding of Python even more. I know I've repeated the point ad nauseam, but you'll thank me for focusing on Python so much here once you start building Raspberry Pi projects.

## CHAPTER 11

Programming Raspberry Pi with Python— Next Steps

By the time you've had the chance to study the material in Chapter 10, "Programming Raspberry Pi with Python—Beginnings," you should have (at the least) the following Python skills under your belt:

- You understand a bit of the purpose behind the Python programming language and why the Raspberry Pi Foundation wanted it to serve as the fundamental development environment on the Pi.
- You know how to get in and out of the Python 3 interpreter and get online help for command syntax.
- You know how to run .py Python script files.

My learning goals for you in this chapter are as follows:

- To understand how to use the IDLE environment
- To have a basic understanding of Python command syntax
- To know where to go to learn Python formally, from "soup to nuts"
- To understand what modules are and how to import them into Python 3

The skills you pick up in this chapter are especially important because when you start building Raspberry Pi projects, you need to understand how to manage Python modules and scripts as well as understand how the code flow works.

I finish this chapter by giving you some pointers for additional resources you can turn to if you're inspired to deep-dive into Python. Let's get to work!

Getting Comfortable with IDLE

Fire up Raspbian and double-click the IDLE3 icon on the LXDE desktop. You'll see the Python Shell open onscreen, as shown in Figure 11.1. The Python Shell is essentially the Python 3 interpreter with a bunch of integrated development environment (IDE) stuff like debugging tools built-in. Interestingly, IDLE is itself a Python application!

NOTE: WHY PYTHON 2?

You've doubtless noticed that Raspbian includes both Python 2 and Python 3 and that a separate version of IDLE exists for each language version. In my opinion, Python 2 is included in Raspbian primarily for backward compatibility with older Python scripts. In fact, the sample games that are included in the Raspbian image are themselves Python 2 scripts.

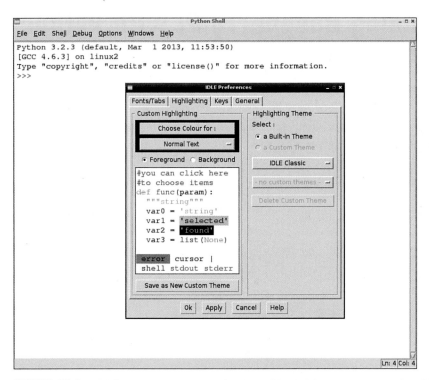

FIGURE 11.1 IDLE, also called the Python Shell. You can open the IDLE Preferences dialog by clicking Options, Configure IDLE.

Let me briefly explain the purpose of each menu in the IDLE Python Shell:

- **File**: Used to create and manage .py Python script files.
- **Edit**: Enables you perform typical word-processing functions (Python scripts are plain text files, after all).
- **Shell**: Allows you to restart the Python Shell if something goes wrong (akin to rebooting a frozen computer).

■ **Debug**: Enables you access tools for troubleshooting your Python scripts.

■ **Options**: Enables you customize the IDLE environment to suit your tastes.

■ **Windows**: Lets you switch among several open script files and the Python Shell.

■ **Help**: Gives you access to the IDLE and Python documentation.

We're going to get right into Python development, here. Try issuing the following statements directly into IDLE at the chevron (>>>) prompt. Remember to press Enter after typing each statement:

```
25*5
```

The asterisk represents multiplication. Try division (/), addition (+), and subtraction (-) as well.

```
len("python")
```

The len function reports on how many characters a given string consists of.

```
x = input("What is your name? ")
```

You are storing user input in a new variable named x. I added a space intentionally after the question mark to put some space between the prompt and the user response.

```
x
```

You can persist the value of a variable in the current Python Shell session. If you restart the shell, then the variable is destroyed.

```
print("I'm gonna add a new line underneath this text. \n")
```

This command uses the newline escape character (**\n**) to add an extra line—this makes your programs easier to read.

```
#This is a comment
```

Single-line comments are preceded with the octothorpe or pound sign (#) character.

```
mylist = ["item one" , 2, 3.14]
```

Lists are a great way to pack multiple pieces of data into a single variable.

```
print(mylist)
```

You can retrieve individual items from a list as well.

```
type(x)
```

The type function tells you what data type is associated with a particular variable.

In preparation for your second real Python script (you wrote your first one in Chapter 10), let's create a new file and save it to your home directory.

TASK: CREATING A NEW PYTHON SCRIPT FILE

You already know that "the journey of a thousand miles starts with the first step." Likewise, before you can author the code in a Python script file, you need to create said script file in the first place. Let's get this party started!

1. In Python Shell, click File, New Window.

2. In the Untitled window that appears, click File, Save.

3. In the Save As dialog box, note that the default save location is your home directory. Name the new file **guessing_game** and click Save.

4. As a test, click File, Open in the editor window. Verify that guessing_game.py exists in your home directory.

You now know how to open script files in the Python Shell!

Writing a Simple Game

Next we are going to write a simple number-guessing game that gives you the opportunity to practice with some common Python code constructions and perhaps have a bit of fun in the process.

Start with the guessing_game.py file you created in the preceding section. Take a look at the following code sample (don't include the line numbers) and then follow that up by studying my annotations for each line of code.

For reference, check out Figure 11.2 to see what the completed script looks like on my Raspberry Pi.

```
1. /usr/bin/env python
2. #Number guessing game adapted from
   #inventwithpython.com
3. import random
4. guesscount = 0
5. number = random.randint(1, 10)
6. print("I thought of a number between 1 and 10. Can you guess it in three
   tries?\n")
7. while guesscount < 3:
8.    guess = input()
9.    guess = int(guess)
10.   guesscount = guesscount + 1
```

11.   if guess < number:
             print("Too low.")

12.   if guess > number:
             print("Too high.")

13   if guess == number:
             break

14. if guess == number:

15.   guesscount = str(guesscount)

16.   print("Congratulations! You guessed the correct number in " +
guesscount+ " guesses!")

17. if guess != number:
             number = str(number)
             print("Sorry. The number I thought of was " + number + ".")

```
#!/usr/bin/env python

# Number guessing game adapted from
# inventwithpython.com

import random

guesscount = 0

number = random.randint(1, 10)

print("I thought of a number between 1 and 10. Can you guess it in three tries?\n")

while guesscount < 3:
    guess = input()
    guess = int(guess)

    guesscount = guesscount + 1

    if guess < number:
        print("Too low.")

    if guess > number:
        print("Too high.")

    if guess == number:
        break

if guess == number:
    guesscount = str(guesscount)
    print("Congratulations! You guessed the correct number in " + guesscount + " guesses!")

if guess != number:
    number = str(number)
    print("Sorry. The number I thought of was " + number + ".")
```

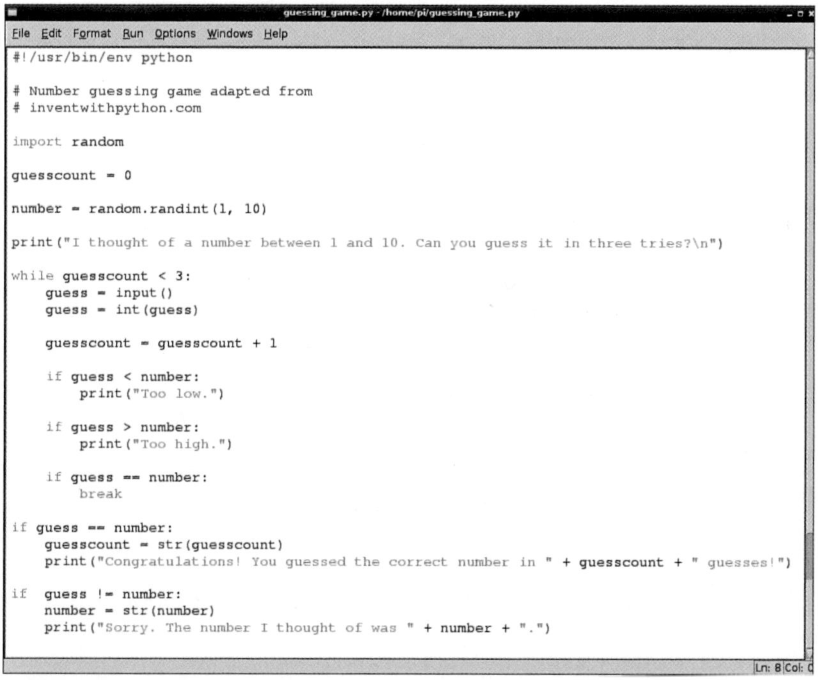

FIGURE 11.2   The number guessing game source code

On to the purpose of each line in the program:

1. This is the "shebang" line that points Raspbian to the location of the Python interpreter.

2. These are two single-line comments that give credit to the developer on whose code this example is based on. Incidentally, multiline comments in Python are done using the triple quote (""") punctuation before and after the comment.

3. Use the import function to bring in external code modules into the Python environment. Modules are discussed in greater detail later in the chapter. For now, understand that random is a module that ships with Python and gives you access to functions related to (what else?) random number generation.

4. Create a variable to store the running count of user guesses and initialize the value of the variable to zero.

5. Define a variable to hold the randomly selected number. Specifically, you call the randint function inside of the random module and ask the Python interpreter to generate an integer (whole number) between 1 and 10, inclusive.

6. This print statement explains the game to the player and inserts a new line between this prompt and the user's first guess.

7. The while statement is an example of *looping logic*. It says "keep repeating whatever code is indented underneath until the guesscount variable reaches 3."

8. Populate the guess variable with the user's typed response.

9. Use the int function to ensure that the user's input is typed as an integer. This is an example of type casting, in which you can convert data from one type to another.

10. Increment the guesscount variable by one each time you loop through the indented while code.

11. The if statement is probably the most common looping function in Python. Here it tests the guess variable against the computer's generated number. If the user's guess is below the number, it tells the user.

12. This if block does the same thing as 11, but here it tests if the user's guess is above the correct value.

13. If the user's guess matches the computer's randomly selected number, then you break out of the while loop and continue with whatever code comes next in the script.

14. This if statement (and the next one) are necessary because if you break out of the while loop with a correct response, you want to end the game. This line of code also uses concatenation to combine static text and variable data. More on that later.

15. Convert the guesscount variable, which was created as an integer, to a string value. It's common practice to cast numbers to strings when you want to print output for the user.

16. Concatenate, or combine, static text and variable data using the plus (+) operator. This can get confusing because you can also use the plus sign to perform arithmetic addition.

17. The purpose of this block is to handle the situation in which you leave the while loop because the user's guess count exceeds three tries. Here you verify that the user's guess does not match the computer's number (!= is the programmatic equivalent of "not equal to"), convert the number to a string, and then inform the user.

Delving into a Bit More Detail

You can close your Python script file; let's work directly in the Python Shell. First, I want to discuss three Python programming features in a bit more detail:

- variables
- type casting
- concatenation

I'm calling out these three programming tools because they are so fundamental not only to Python, but to any programming language. For instance, most computer programs take data and perform some sort of processing and evaluation on it. How and where do you store that data? What if you need to convert data from one form to another—how is that done in Python? Finally, how do you combine multiple pieces of dynamic data?

Read on, friend...read on.

Variables

As previously discussed, a *variable* is a named placeholder for data. Variable naming in Python 3 is flexible, but there are a few rules that you need to keep in mind:

- Python key words cannot be used as a variable name (naming a variable print is not allowed, for instance).
- Variable names cannot contain spaces (underscores are okay, though).
- Uppercase and lowercase characters are distinct (Var1 and var1 are considered two separate variables).
- The first character of a variable name must be a letter a through z, A through Z, or an underscore character (no numbers to start variable names because this confuses the Python interpreter).
- After the first character, you can use the digits 0 through 9 and underscores in variable names.

The equal sign (=) is used to assign value to a variable. This is in stark contrast to the double equals (==) that are used to test equality between two values. For instance:

- **var1 = 2** : This statement says, "The value of the variable named var2 is 2."
- **var1 == 2** : This asks the question, "Does the value of the variable named var1 equal 2?"

Type Casting

In programming, a variable needs to be associated with a data type. The data type constrains, or limits, the kind of data stored by the variable. For instance, does the variable x below store a number or a string of characters? How about the variable y? How do you think Python computes the result of the variable z in the third code example?

```
x = "234"
y = 432
z = x + y
```

In some programming languages, the variable x in the previous example would be assumed to be a string because of the quotation marks. Therefore, the expression x+ y would fail because you can't add a string and a number together.

Strictly typed programming languages like C require that you declare not only a variable's name, but also the type of data that it can hold. Python isn't like that; it's much more lax.

Yes indeed—Python is pretty forgiving, data type-wise. You can use the type function to check the data type that Python associated with a variable. Try the following:

```
type(x)
```

Python 3 supports the following native data types:

- **Boolean**: Possible values are True or False.
- **Numbers**: Integers (whole numbers); Floats (decimal or fractional numbers), or complex numbers.
- **Strings**: Character sequences.
- **Bytes**: Binary data such as images or other media files.
- **Lists**: Ordered value sequences.
- **Tuples**: Ordered values that are different from lists inasmuch as lists can change their values (mutable), but tuples cannot (immutable).
- **Sets**: Unordered value sequences.
- **Dictionaries**: Unordered key-value pairs.

You can use type-casting functions to manually convert data from one data type to another. This is useful when you want to ensure that Python receives variable data in a particular format.

For instance, try this:

```
vara = "100"
type(vara)
```

The result of the above code is that Python sees "100" as a string rather than as an integer. Does that result surprise you? It shouldn't. When the Python interpreter saw data contained within quotes, it assumed you wanted to use the str (string) data type instead of int (integer). You can fix it, though, by converting the string you assigned to the variable to an integer:

```
varb = int(vara)
type(varb)
```

In other words, Python infers a data type based upon how you type. If you use quotes, then dollars to donuts Python assumes you're talking about a string value. If you supply a number without quotes, then Python selects one of the numeric data types depending upon the number.

For instance, a variable value of 100 is an integer, and 100.1 is a float. Understand?

Concatenation

In Python, string concatenation enables you to patch together bits of code. To do this, simply use the plus sign (+). Consider the following examples, all of which you can try out in the Python Shell:

- **str1, str2 = "abc", "def"**: Two things—you can create and initialize more than one variable in one shot, and you can use single or double quotes to contain string data.

- **str1 + str2**: Note that when you concatenate strings, no extra space is included. The result of this operation is abcdef.

- **str1 + " " + str2**: You can pad spaces by passing in the space character(s) as a separate string literal. In this example, the result is "abc def". Pretty cool, huh?

- **print("The combined string value is: " + str1 + str2)**: You can concatenate string literal data with variable data to provide the user with customized output.

Modules

In Python, modules are .py script files that contain one or more related code blocks. What is so cool about modules is that they make programming much more modular. Think of it: Would you rather write (or copy/paste) a bunch of functions you created that pertain to several different Python programs you're working on, or would you rather have those functions stored in a module that you can load and unload at your convenience?

Earlier in this chapter I briefly introduced the random module that ships with the so-called reference version of Python 3. As you can see in Figure 11.3, the contents of the module are stored in plain text and can be viewed and analyzed by anyone.

```
                      random.py - /usr/lib/python3.2/random.py                    _ □ x
File  Edit  Format  Run  Options  Windows  Help

    def randint(self, a, b):
        """Return random integer in range [a, b], including both end points.
        """

        return self.randrange(a, b+1)

    def _randbelow(self, n, int=int, maxsize=1<<BPF, type=type,
                   Method=_MethodType, BuiltinMethod=_BuiltinMethodType):
        "Return a random int in the range [0,n).  Raises ValueError if n==0."

        getrandbits = self.getrandbits
        # Only call self.getrandbits if the original random() builtin method
        # has not been overridden or if a new getrandbits() was supplied.
        if type(self.random) is BuiltinMethod or type(getrandbits) is Method:
            k = n.bit_length()  # don't use (n-1) here because n can be 1
            r = getrandbits(k)          # 0 <= r < 2**k
            while r >= n:
                r = getrandbits(k)
            return r
        # There's an overriden random() method but no new getrandbits() method,
        # so we can only use random() from here.
        random = self.random
        if n >= maxsize:
            _warn("Underlying random() generator does not supply \n"
                  "enough bits to choose from a population range this large.\n"
                  "To remove the range limitation, add a getrandbits() method.")
            return int(random() * n)
        rem = maxsize % n
        limit = (maxsize - rem) / maxsize   # int(limit * maxsize) % n == 0
        r = random()
        while r >= limit:
                                                                Ln: 216 Col: 63
```

FIGURE 11.3   Python modules are note encrypted, but boy, are they useful! Here you can see the code behind the randint function used earlier in the chapter.

NOTE: WHERE ARE MODULES LOCATED?

You can run help("modules") to get a list of all currently available modules in your current Python 3 installation. After you get the name of your desired module, type **help("module_name")** to get the file location. For instance, in Raspbian, the random module is located by default in /usr/lib/python3.2/random.py.

Assuming a module is present on your system (see the note "Where Are Modules Located?" for more info), you can use the import statement to bring a module into the current Python environment. Note that an import statement lasts only for the duration of the current script file.

When you perform an import, all of the code contained in the module becomes available to you in Python. For instance, to import all code from the math module, you can issue any of the following statements:

```
import math

from math import *

import math, random (we can import more than one module at a time; just use a
comma separator)
```

After you've imported a module, run dir(module_name) to get a list of all the names (the Python term for code components) that are contained inside the module. To illustrate, run the following three statements in the Python Shell:

```
import math
content = dir(math)
content
```

Now let's drill into the math module, and you'll see how to take advantage of a module's inner content. As an example, let's work with the sqrt function from the math module:

```
import math
math.sqrt(25)
```

With respect to Python programming, a *fully qualified* function name takes the form of module.function. Thus, after importing the math module, you issue math.sqrt() when you want to run the sqrt() function that is contained in the math module.

Even though you imported the math module, the Python interpreter would get confused and issue an error if you used just sqrt() in your code without qualifying its location.

NOTE: WHERE TO FIND COOL MODULES?

I've found that you can learn about any Python 3 module directly from the Python website. Check out the Python Module Index (http://is.gd/yr7nOA) to learn about the built-in module library. For third-party modules, see the Useful Modules list (http://is.gd/OvwCJm) at the Python Wiki. Finally, I cover Raspberry Pi-specific modules as we move through the remainder of this book.

As I said earlier, many Raspberry Pi projects require that you obtain and install additional modules. You can use the Linux apt-get command in many cases.

One word of warning: You need to be mindful of the fact that you're working with Python 3 and not Python 2. Many online tutorials show you how to do stuff with Python on the Pi, and the module and code references the older version of Python.

Let's make sure you have the most recent version of the GPIO module in your Python 3 installation. This module is important later because, you'll recall, the GPIO headers are the principal way that you connect the Raspberry Pi to external hardware.

I've found that the case-sensitivity in Linux has caused Raspberry Pi users to conclude that their Python 3 installation is missing certain modules when, in point of fact, they are present. Try the following procedure.

TASK: LOADING AND THEN UPDATING THE GPIO MODULE IN PYTHON 3

Many of the projects that I cover in the latter part of this book involve taking control of the Raspberry Pi's General Purpose Input/Output (GPIO) header pins. Accordingly, it is crucial that you ensure that your Python installation has access to the GPIO modules.

1. From LXTerminal, type **python3** to start an interactive Python 3 session.

2. Import the GPIO module included in Raspbian so you can begin the process of interacting with the Pi's GPIO headers:

```
import RPI.GPIO
```

3. Did that work? No? Well, something you should know is that GPIO is a function library inside of the RPi module. Notice the mixed case. Try this:

```
import RPi.GPIO as GPIO
```

The as keyword is used to provide an alias to an imported module. This means you can call GPIO functions by using GPIO instead of RPi.GPIO. You had some more problems though, correct? It turns out you also need to run Python as root. Sheesh!

4. Run **exit()** to leave the interpreter and then issue **sudo python3** to enter the interpreter as root. One more time with feeling!

```
import RPi.GPIO as GPIO
dir(GPIO)
```

Now we're cooking! The output is displayed in Figure 11.4.

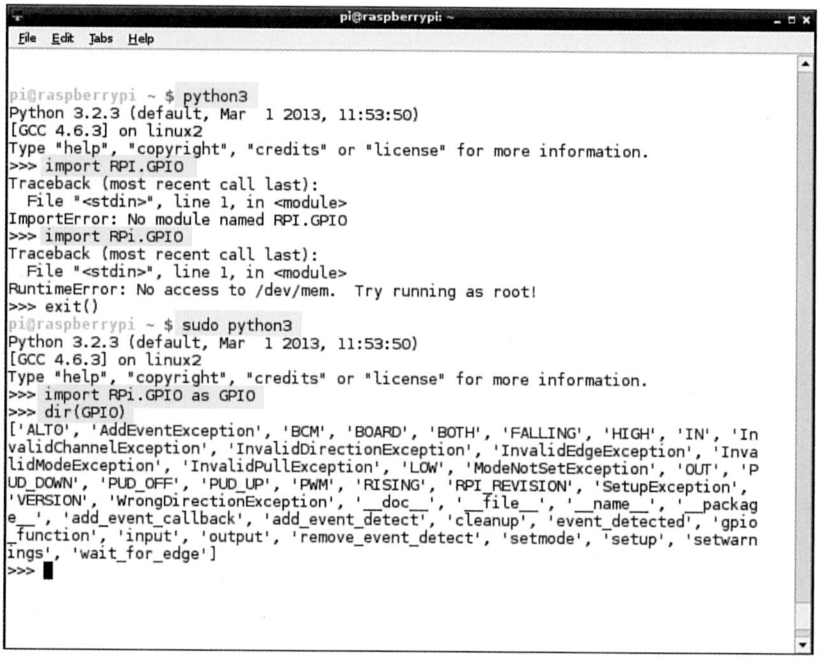

```
                                    pi@raspberrypi: ~                          _ □ x
File  Edit  Tabs  Help

pi@raspberrypi ~ $ python3
Python 3.2.3 (default, Mar  1 2013, 11:53:50)
[GCC 4.6.3] on linux2
Type "help", "copyright", "credits" or "license" for more information.
>>> import RPI.GPIO
Traceback (most recent call last):
  File "<stdin>", line 1, in <module>
ImportError: No module named RPI.GPIO
>>> import RPi.GPIO
Traceback (most recent call last):
  File "<stdin>", line 1, in <module>
RuntimeError: No access to /dev/mem.  Try running as root!
>>> exit()
pi@raspberrypi ~ $ sudo python3
Python 3.2.3 (default, Mar  1 2013, 11:53:50)
[GCC 4.6.3] on linux2
Type "help", "copyright", "credits" or "license" for more information.
>>> import RPi.GPIO as GPIO
>>> dir(GPIO)
['ALTO', 'AddEventException', 'BCM', 'BOARD', 'BOTH', 'FALLING', 'HIGH', 'IN', 'In
validChannelException', 'InvalidDirectionException', 'InvalidEdgeException', 'Inva
lidModeException', 'InvalidPullException', 'LOW', 'ModeNotSetException', 'OUT', 'P
UD_DOWN', 'PUD_OFF', 'PUD_UP', 'PWM', 'RISING', 'RPI_REVISION', 'SetupException',
'VERSION', 'WrongDirectionException', '__doc__', '__file__', '__name__', '__packag
e__', 'add_event_callback', 'add_event_detect', 'cleanup', 'event_detected', 'gpio
_function', 'input', 'output', 'remove_event_detect', 'setmode', 'setup', 'setwarn
ings', 'wait_for_edge']
>>> █
```

FIGURE 11.4 Working with modules in Python 3 can be...interesting.

5. Exit the interpreter one final time. Let's update the module to make sure you have the latest and greatest version:

```
sudo apt-get update
sudo apt-get dist-upgrade
sudo apt-get install python3-rpi.gpio
```

Where Do You Go from Here?

If nothing else, I hope your work in Chapters 10 and 11 has fired your imagination and inspired you to learn more about Python programming. My challenge as your guide has been to pack as much Python instruction as possible in just a few pages in a Raspberry Pi book.

However, for those interested readers, I want to share with you what I think are the very best Python learning resources available. I know different people have different learning styles, so following is a collection of various types of references for your studying pleasure.

- **Textbooks**: For my money, you simply cannot go wrong with Tony Gaddis' *Starting Out with Python, 2nd Edition* (http://is.gd/CZy0QN).

  Another Python text I enthusiastically recommend is Mark Lutz' *Learning Python* (http://is.gd/0oueEV). I'm not sure why the book gets mixed reviews on Amazon because it really is a landmark text.

- **Computer-based Training**: At the risk of coming across as a self-promoter, I recorded a computer-based training course on Python Programming for CBT Nuggets (http://is.gd/A5XQei) that I fully stand behind. What's cool about computer-based training is that you can see the concepts in action immediately on your computer screen.

  A second computer-based training course I had a hand in developing and recommend is Wesley Chun's *Python Fundamentals LiveLessons* (http://is.gd/V56Ekl).

- **Online Resources**: As I've stated before, the Python website is perhaps the best reference source on the Internet for learning Python. Check out The Python Tutorial at http://is.gd/KyCom5.

  Another awesome online resource, and it is completely free, is *Dive into Python 3* by Mark Pilgrim (http://is.gd/QeW7OH). This is essentially the full text of the associated textbook by Apress. However, it's really nice to have direct access to the source code and examples.

## CHAPTER 12

Raspberry Pi Media Center

I am a huge set-top box fan. Not the rubbish one your cable company gave you, but the ones that let you connect to virtually every streaming service under the sun. In fact, if it weren't for my wife and daughter, I would have gotten rid of digital cable service years ago and devoted myself entirely to streaming media services such as Netflix and Hulu. The only way that somebody will take my Apple TVs away from me is from my cold, dead hands!

One of the biggest attractions of the Raspberry Pi is its utility as a media center platform. Recall that the combination of the Broadcom VideoCore IV GPU and the HDMI output means that you have the capability of sending 1080p High Definition to your monitor or HD television.

As it always happens in open source development, several software options exist for building a Raspberry PI media center. Some of the most popular choices include

- **Raspbmc**: http://is.gd/OX7dMY
- **RasPlex**: http://is.gd/HfEIIi
- **OpenELEC**: http://is.gd/drMs1E
- **Xbian**: http://is.gd/7LUXtc

Three of these software/OS packages are derivatives of XBMC Media Center (http://is.gd/xwVddv), the gold standard in open source media player software. You can see a screen shot of the XBMC interface in Figure 12.1.

FIGURE 12.1 XBMC is the gold standard in open source media center software.

Note that because I wanted the highest-quality screen shots for this book, the XBMC interface images I give you in this chapter are from the Windows version rather than from Raspbmc. The good news is that XBMC 12.2 "Frodo" looks and behaves exactly the same regardless of its host hardware—hence the great beauty of platform-independent, open source software.

A Bit o' History

The name XBMC originally stood for Xbox Media Center because the software was intended to run only on modified ("modded") Xbox consoles. These days, of course, XBMC runs on almost every desktop or mobile platform, and to that end, uses the backronym "XBMC Media Center."

NOTE: XBMC AND RASPBIAN

For those who are more experimentally minded, you can actually install XBMC directly on top of Raspbian. Check out the Raspbian XBMC project page (http://is.gd/vxKwGJ) for more details.

I like to describe XBMC as a Swiss Army knife media application that can play just about any media file you can throw at it. Here is a run-down of some of the best features of XBMC:

- **Plugins**: You can easily extend XBMC functionality by installing add-ons that (for instance) display local weather, pull Internet Movie Database (IMDB) metadata for your currently loaded media, and so forth.
- **Media Scrapers**: XBMC can scan any media you load and automatically detect everything there is to know about the item. For instance, the scraper can detect an MP3 audio file's album track listing, song lyrics, and so on.
- **Apps**: You can launch applications that enable XBMC to tie into streaming media services such as Netflix.
- **Codec Richness**: Compressor-Decompressors (Codecs) allow media player software to recognize, decode, and play various media. Not only does XBMC ship with a ton of media codecs, you can manually install additional codecs to ensure that your custom media is playable from within XBMC.

For a more comprehensive listing of XBMC features, read the associated Wikipedia article at http://is.gd/yZmSK8.

But Will It Blend?

If you haven't yet seen the Blendtec's "Will It Blend?" viral marketing videos, then do yourself a favor and check them out on YouTube at http://is.gd/8gWfN9.

"Will It Blend?" reminds me of a question that is more cogent to us as Raspberry Pi enthusiasts—namely, "But does the Raspberry Pi have enough processing power to run XBMC appropriately?"

Ah yes, the eternal question. First things first: The Model A board simply will not do as a media player. Number one, there is the limited memory issue. Number two, you'll need a hardwire Ethernet connection to get appreciable network speeds, and the Model A has no RJ-45 port.

Yes, yes—I know what you are thinking: "Couldn't I add an RJ-45 wired Ethernet port to my Model A Pi by using USB?" This is true enough, but I nonetheless submit that the latency you'll experience in not employing wired Ethernet that is built into the Pi's circuitry (like we have with the Model B) makes the process barely worthwhile.

As in all things, though, your mileage may vary.

The primary differentiator among the various Raspberry Pi XBMC ports is how completely they take over the host operating system. Remember that with a device as hardware-constrained as the Pi, the fewer software layers you have between XBMC and the underlying hardware, the better.

There is vociferous debate online as to who makes the best XBMC player for the Pi. However, in this book we install Raspbmc because it is (generally) considered to be the most stable and mature XBMC port.

Introducing Raspbmc

As you probably guessed, Raspbmc (http://is.gd/OX7dMY) is a portmanteau of *Raspbian* and *XBMC*. Thus, Raspbmc replaces the operating system on your Raspberry Pi instead of serving as a third-party app that you install on an existing Raspbian installation.

Raspbmc is the brainchild of Sam Nazarko of London and achieved final 1.0 release status in February 2013. For in-depth coverage of all things Raspbmc-related, see Sam's book *Raspberry Pi Media Center*, by Packt Publishing (http://is.gd/GDTsVR).

Basically, Sam stripped Raspbian down to almost bare metal and wove XBMC on top of it. To that point, don't expect to get anything more than Terminal access to your Raspbmc system. Because X11 is not present, you'll never get VNC remote connections to work.

You need the following components to get Raspbmc up and running on your Raspberry Pi Model B:

- **HDMI connection**: You need a monitor or television (remember, no VNC support in Raspbmc).
- **A decent-sized SD card**: Go for a 16–32GB, Class 10 card unless you are absolutely committed to storing your media on a USB thumb drive.
- **Keyboard and mouse**: These peripherals are necessary due to the "no VNC" rule mentioned previously.
- **Wired Ethernet connection with DHCP**: Although you can get Wi-Fi going easily enough after Raspbmc is installed on your Pi, your best bet—at least during the installation phase—is to plug in a physical Ethernet cable and let the Pi pick up an IP address from your local DHCP server.

TASK: INSTALLING RASPBMC UNDER WINDOWS

1. Download the Windows Raspbmc Installer from the Raspbmc website (http://is.gd/jdMxS2) and extract the .ZIP file contents to a local directory on your Windows system.

2. Open Setup.exe, which starts the Raspbmc Installer. The interface is shown in Figure 12.2.

FIGURE 12.2 You can flash your SD card under Windows by using the Raspbmc Installer application.

3. Take pains to verify that you select the correct volume in the device list. You want to flash an SD card, remember, and you don't want to pick the wrong drive. (You also need to select the *I accept the license agreement* option before clicking Install to flash your card.)

4. When the installation process completes, you see a *Congratulations!* message informing you to boot your Raspberry Pi from the newly flashed SD card. Note also that you need to plug in the Ethernet cable into your Model B board, have your home network configured for automatic (DHCP-based) IP address assignment (if it's not already), and have your keyboard, mouse, and HDMI monitor all plugged in and ready to rock.

TASK: INSTALLING RASPBMC UNDER OS X

1. Open up a Terminal prompt and download the Python 2-based installer:

```
sudo curl -O http://svn.stmlabs.com/svn/raspbmc/testing/installers/python/
install.py
```

NOTE: CURL

The curl program is used in Linux to fetch web-based content from a command line. Note also that the parameter after curl is an O (the letter), as opposed to a zero.

w run the install.py Python script:

```
sudo python install.py
```

Hey, you are applying some of our newly found Python skills already—awesome! You can view the screen output of these commands in Figure 12.3.

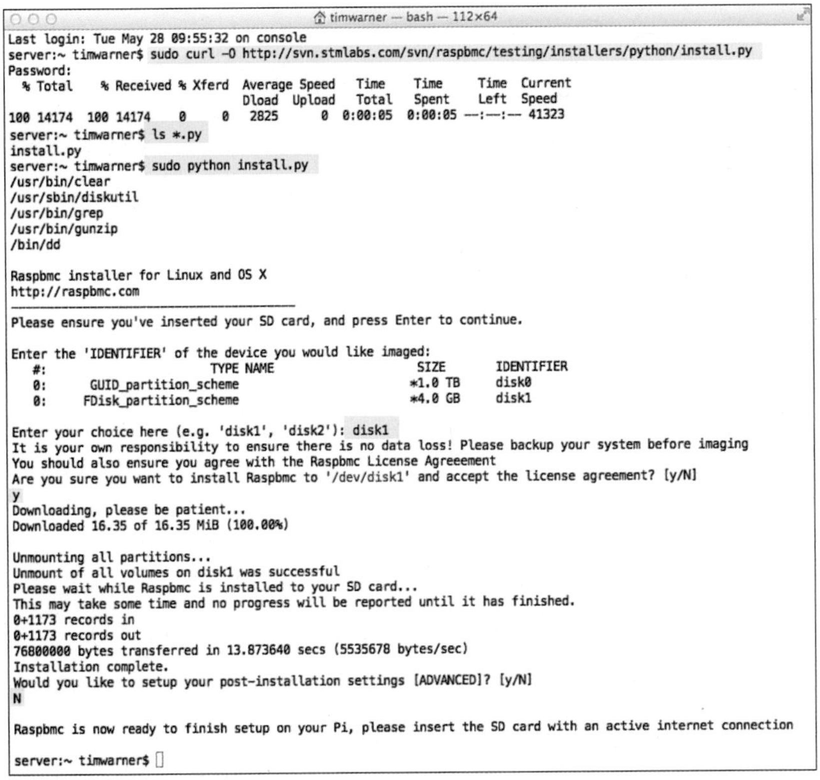

```
● ● ●                          ☆ timwarner — bash — 112×64
Last login: Tue May 28 09:55:32 on console
server:~ timwarner$ sudo curl -O http://svn.stmlabs.com/svn/raspbmc/testing/installers/python/install.py
Password:
  % Total    % Received % Xferd  Average Speed   Time    Time     Time  Current
                                 Dload  Upload   Total   Spent    Left  Speed
100 14174  100 14174    0     0   2825      0  0:00:05  0:00:05 --:--:-- 41323
server:~ timwarner$ ls *.py
install.py
server:~ timwarner$ sudo python install.py
/usr/bin/clear
/usr/sbin/diskutil
/usr/bin/grep
/usr/bin/gunzip
/bin/dd

Raspbmc installer for Linux and OS X
http://raspbmc.com
─────────────────────────────
Please ensure you've inserted your SD card, and press Enter to continue.

Enter the 'IDENTIFIER' of the device you would like imaged:
   #:                     TYPE NAME           SIZE       IDENTIFIER
   0:      GUID_partition_scheme             *1.0 TB     disk0
   0:      FDisk_partition_scheme            *4.0 GB     disk1

Enter your choice here (e.g. 'disk1', 'disk2'): disk1
It is your own responsibility to ensure there is no data loss! Please backup your system before imaging
You should also ensure you agree with the Raspbmc License Agreeement
Are you sure you want to install Raspbmc to '/dev/disk1' and accept the license agreement? [y/N]
y
Downloading, please be patient...
Downloaded 16.35 of 16.35 MiB (100.00%)

Unmounting all partitions...
Unmount of all volumes on disk1 was successful
Please wait while Raspbmc is installed to your SD card...
This may take some time and no progress will be reported until it has finished.
0+1173 records in
0+1173 records out
76800000 bytes transferred in 13.873640 secs (5535678 bytes/sec)
Installation complete.
Would you like to setup your post-installation settings [ADVANCED]? [y/N]
N

Raspbmc is now ready to finish setup on your Pi, please insert the SD card with an active internet connection

server:~ timwarner$ []
```

FIGURE 12.3 Flashing your Raspbmc SD card under OS X isn't as easy as it is under Windows.

Configuring Raspbmc

Flashing your SD card by using the Raspbmc installer does not actually fully set up your environment; instead, the installer formats your SD card, creates one big partition, and installs the barest layer of Raspbian on the card.

Upon first bootup the Raspbmc installer takes over the system, downloading and applying Raspbian to your Pi. You can see this in action in Figure 12.4.

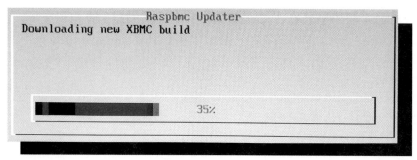

FIGURE 12.4 Raspbmc automatically installs the latest version of the software during the Pi's first startup.

After installation completes, you see the XBMC interface and are asked to choose a default language. You need to navigate to System, Settings, Appearance, International to verify and set your localization settings for

- Region
- Character Set
- Timezone country
- Timezone

You'll find that XBMC navigation is pretty intuitive by using the keyboard and mouse. As shown in Figure 12.5, each configuration window can be closed individually by clicking the "X" in the upper-right corner of each window. Also the navigation buttons in the lower-right of each screen take you back one screen or jet you to the Home screen, respectively.

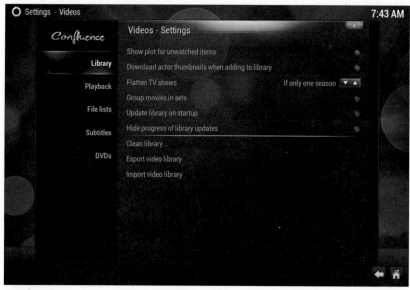

FIGURE 12.5 The XBMC user interface can be navigated easily with a keyboard, mouse, or infrared (IR) remote.

Getting Your Remote Control Working

What is a set-top box worth if you can't control it with your favorite remote control? Nothing, in my estimation. Your first order of business is to study the list of Raspbmc-compliant remotes at the Raspbmc website (http://is.gd/5Sw23o).

Second, you need to decide whether you want to control your Raspbmc box via Internet Protocol (IP) or Infrared (IR). An example of an IP-powered remote control is an iOS or Android app that enables you to control your Raspbmc media center.

IR is a line-of-site remote control protocol that has been around seemingly forever; I'm sure that you use IR remotes to control your television sets right now.

If you decide to go the IP remote control route, then you should investigate mobile apps designed to control XBMC over your local IP network:

- **Official XBMC Remote (Android)**: http://is.gd/k4UeWY
- **Official XBMC Remote (iOS)**: http://is.gd/qZkt3l

Just for grins, let me show you how to set up your iOS-based XBMC Remote app to connect to your Raspbmc system.

TASK: CONTROLLING YOUR RASPBMC BOX FROM IOS

1. In XBMC, navigate to System, Settings, Services, Remote Control and enable Allow programs on other systems to control XBMC.

2. On the same configuration page, navigate to Zeroconf and enable Announce these services to other systems via Zeroconf. For more information on Zeroconf, see the note "What is Zeroconf?"

3. One more setting group: On the Webserver page, ensure that Allow Control of XBMC via HTTP is enabled and optionally change the listener port (8080 is a good choice) and add a username and password (xbmc/xbmc is a common combination).

4. Verify your IP address by going back to the home page, navigating to System, System Info, and checking out the IP address field.

5. Download and install the Official XBMC Remote from the App Store.

6. Start the app and tap Add Host.

7. In the New XBMC Server dialog box, add as many details as you can regarding your Raspbmc (see Figure 12.6).

NOTE: WHAT IS ZEROCONF?

Zeroconf is a shorthand notation for *Zero-Configuration Networking*, which is a collection of technologies that operates over TCP/IP and allows network devices such as computers and mobile hardware to communicate without the need for special setup procedures. For instance, Apple has a Zeroconf protocol called Bonjour that enables, for instance, your iPhone to discover your iMac's iTunes music library and stream the songs from the computer to the mobile device. Pretty neat, eh?

FIGURE 12.6   The more information you can provide to the remote control app, the better the chance is that the app will discover your Raspbmc box on your network.

8. Tap Save to save your configuration, and tap Find XBMC to locate your device on the network.

With a successful connection, you can fully control your Raspbmc box remotely as shown in Figures 12.7 and 12.8.

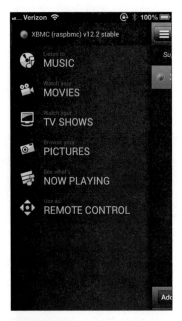

FIGURE 12.7 How cool is it that you can manage your XBMC library from your mobile device?!

FIGURE 12.8 The Official XBMC Remote app serves as a, well, remote control for your Pi (among many other things).

TASK: CONTROLLING YOUR RASPBMC BOX FROM A WEB BROWSER

1. Ensure that Raspbmc is configured to allow HTTP access as outlined in the previous procedure.

2. Fire up a web browser from a computer on the same LAN as your Raspbmc system and navigate to the proper URL. You can obtain your system's IP address by navigating from the home page to System, System Info and examining the IP address field.

3. Next, navigate to System, Settings, Services, Webserver and enable the option Allow control of XBMC via HTTP. For instance, imagine my system has the IP address 10.1.10.1. Now I can open a web browser, type http://10.1.10.1 as the address, and remotely connect to my Raspbmc server.

4. Upon a successful connection, you are presented with a simple, yet intuitive method for managing the content on your Raspbmc device. The default web interface is shown in Figure 12.9.

NOTE: EXTENSIBILITY, HO!

In the XBMC Webserver Property page, note that you can download additional web interface skins that change the website's look and feel. Click Default and then click Get More... to do some shopping. Of course, your Raspbmc device must be connected to the Internet for this procedure to work.

FIGURE 12.9 XBMC includes a simple web-based management tool (and remote control, of course).

TASK: CONFIGURING A GPIO-BASED IR RECEIVER

The previous methods work just fine, but what if you want to control Raspbmc by using your hardware remote that operates over infrared (IR)?

If you are thinking like a true (or budding) hardware engineer, you probably thought of the following:

"I'll bet I need to attach an IR sensor to my Pi's GPIO pins to control my Raspbmc box by using a hardware remote control."

If you did think the preceding thought, then kudos to you! Actually, you have two possibilities for IR-based remote communication:

- Purchase an IR receiver and wire it directly to the GPIO pins on the Pi board.

- Purchase a USB IR receiver, plug it into your powered USB hub, and load the appropriate drivers.

Unfortunately, no truly straightforward documentation exists for either procedure as of summer 2013. However, I can give you one way to use the GPIO method by using parts from good ol' trusty Adafruit.

In order to complete this task, you'll need to purchase some items from Adafruit or another parts supplier. The good news is that I'm sure you'll find your efforts duly rewarded—controlling your own custom media center via IR remote ROCKS!

Parts needed:

- **TSOP38238 IR receiver** (http://is.gd/y9XzOC)

- **Female-to-female jumper wires** (http://is.gd/anFo27)

- **Mini remote control** (http://is.gd/dp0mPW)

1. As shown in Figure 12.10, connect three female-to-female jumper wires to (a) the three leads of the IR receiver and (b) three GPIO pins. The receiver-to-GPIO lead mapping is as follows:
    - IR receiver pin #1 to GPIO pin #18 (data)
    - IR receiver pin #2 to GPIO pin GND (ground)
    - IR receiver pin #3 to GPIO pin 3V3 (3.3V power)

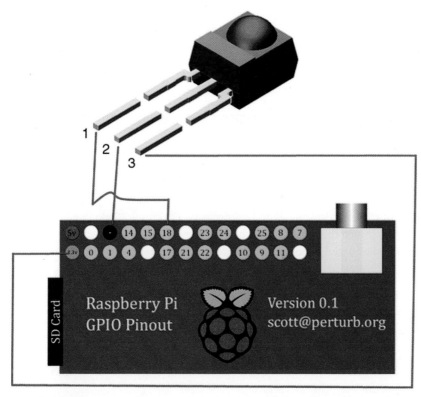

FIGURE 12.10 Schematic showing you how the GPIO pins connect to the IR sensor by means of female-to-female jumper wires.

2. The Linux Infrared Remote Control (LIRC) software component is what orchestrates communication between your hardware IR remote and XBMC. Log into XBMC on your Raspberry Pi and navigate to Programs, Raspbmc Settings, IR Remote.

   While you are testing the hardware remote on its own, ensure that the setting Enable GPIO TSOP IR Receiver is *disabled*. When you're ready to test your hardware remote, come back to this setting and enable it. Remember while you're in the Settings area to make a note of your Raspbmc's IP address.

NOTE: TSOP, WHAT'S THAT?

In electronics, TSOP stands for **T**hin, **S**mall **O**utline **P**ackage and refers to the physical form factor of the IC chip itself. Thus, TSOP is a general descriptive term that applies to many types of IC components, not simply IR receivers.

3. The remainder of the configuration is done from the Linux command line. You can use SSH to connect to your Raspbmc box in the same manner you can with traditional Raspbian. (Remember that you learned how to use SSH in Chapter 7, "Networking Raspberry Pi"). The default username is (surprise, surprise) pi, and the default password is raspberry.

Rather than give you a couple pages of Linux terminal commands, I will simply refer you to the "Using an IR Remote with a Raspberry Pi Media Center" tutorial at Adafruit (http://is.gd/97RvGt).

For instructions and/or advice on getting your own personal IR remote working with Raspbmc, I suggest you turn to the trusty eLinux.org website at http://is.gd/yaJOSw. This site contains a constantly updated list of peripheral devices that have been verified to work with the Raspberry Pi.

Transferring Content to Your Pi

You can upload media files (movies, TV shows, home videos, music, pictures...the list goes on) to your Raspberry Pi using File Transfer Protocol (FTP). I recommend the freeware Filezilla to Windows and Mac users because it is a stable, straightforward FTP client that looks and behaves the same on either OS platform.

You already know the IP address of our Raspbmc box, so you can get right to work.

TASK: UPLOADING MEDIA CONTENT TO RASPBERRY PI

1. Start Filezilla and fill in the details in the Quickconnect bar:

- **Host**: <IP address of your Pi>
- **Username**: pi
- **Password**: raspberry
- **Port**: Leave this field empty because FTP assumes the use of Transmission Control Protocol (TCP) port 21 by default.

2. Connecting as the pi user puts you in the /home/pi home directory by default. Right-click in the Remote site window and click Create Directory to create the following folders for your media:

- Movies
- TV

- Music
- Pictures

You can name your content folders anything you want, of course, but I suggest you avoid spaces (see Figure 12.11).

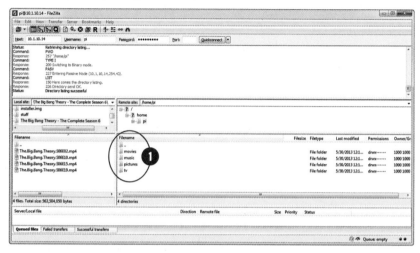

FIGURE 12.11   FileZilla makes it simple to upload your media content to the Raspbmc device.

3. To actually transfer a file, simply drag and drop the file(s) from the OS X Finder, Windows Explorer, or from the Local Site pane in Filezilla over to the appropriate destination in the FileZilla Remote site: pane.

Progress information concerning the uploads can be seen in the bottom pane of FileZilla. Easy as pie!

NOTE: USB IS ALLOWED

If your Raspbmc SD card isn't as large as necessary to store all your stuff, note that you can use a USB thumb drive as a media source. Simply populate the USB stick with your media, plug it into the powered USB hub that is connected to your Pi, and specify the USB location when you configure media detection.

Scraping Your Media

As I mentioned earlier in this chapter, a media scraper is a program that can detect, analyze, and report upon a media file that is present to XBMC. For instance, you can upload a season of *The Big Bang Theory* and let the built-in media scrapers fill in metadata such as plot summary, cast, original air date, and so forth.

The types of media that scrapers can detect are as follows:

- Video games
- Music
- Movies
- TV shows
- Internet-based videos

Although the media scrapers available in XBMC are excellent, I advise you to name your media files as descriptively as possible (and without spaces) to help them to more efficiently and accurately do their work.

Take a look at Figure 12.12 to see an example of media file naming best practices.

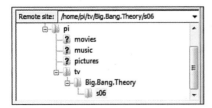

FIGURE 12.12   In order to ensure that media scraping works correctly, individual show files inside my s06 (season 6) folder should be named Big.Bang.Theory. s06e01, Big.Bang.Theory.s06e02, and so on.

The XBMC Wiki has an outstanding article on video library management and file naming best practices that needs to be on your required study list. You can find it at http://is.gd/ RTkODz.

Now that you have your media uploaded and named appropriately, let's configure media detection.

TASK: CONFIGURING XBMC MEDIA DETECTION

In this procedure, you configure media detection for your /home/pi/tv folder.

1. In XBMC, navigate to Video, Files.

2. In the Videos screen, click Files, and then click Add Videos...

3. In the Add Video Source dialog, click Browse and select your target directory. On my system, I want XBMC to detect my legally backed-up TV shows, so I supply the path /home/pi/tv.

4. Optionally add a name for the new media source and then click OK.

5. In the Set Content dialog box, set the This Directory Contains option to your desired media type. In this example, I chose TV shows.

6. In the Choose a Scraper list, select a default choice or click Get More to browse for an alternate choice. You can see the interface in Figure 12.13.

FIGURE 12.13   After pointing Raspbmc to a content location, we instruct the application as to what kind of data the location contains, and which scraper service we want to use.

NOTE: A DESKTOP SCRAPER FOR ANYONE

Windows users can download the free *Media Companion* (http://is.gd/qemjHg) to prepare media file metadata from your personal computer. Then when you transfer the media to your XBMC, the files are prescraped!

When you click OK the selected scraper performs a full scan on your content source. If all goes well, you can go back to the Home screen, click TV shows, and access your library to see all sorts of new artwork and bling to accompany your detected media.

For instance, try right-clicking a television show media file and selecting Episode information from the shortcut menu. The results of selecting information for one of my recorded shows is shown in Figure 12.14.

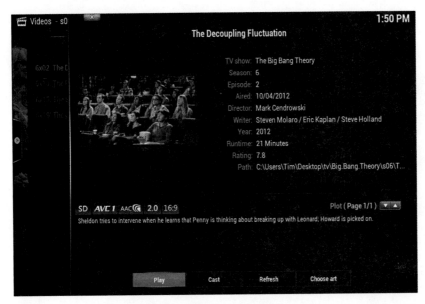

FIGURE 12.14   The ability to provide rich metadata is one of the biggest strengths of XBMC (beyond its ability to play the media, of course!)

Playing Your Content

Let's finish this chapter with a brief tour of the player controls in XBMC. Using Figure 12.15 as your guide, let me explain some of the controls on the XBMC player's onscreen display (OSD):

FIGURE 12.15 You can use your remote control to access the XBMC onscreen display (OSD).

1: These playback controls should be self-explanatory to anybody who has operated a DVD player remote.

2: This control enables you to manage subtitles. For more information on how to use subtitle files with XBMC, see http://is.gd/PmFob0.

3: Here you can tweak video settings, including aspect ratio, black bar cropping, and brightness/contrast.

4: Here you can customize audio settings, including delay, subtitles, and so on.

5: Here you can set, go to, and delete bookmarks. This is one of my favorite features!

Installing Add-Ons

XBMC/Raspbmc add-ons are small apps that extend the functionality of your media player software. You'll find an impressive variety of tools here, giving you the ability to interact with all sorts of cool media, including, but not limited to, the following:

- Apple iTunes podcasts
- Break.com and CollegeHumor prank videos
- Weather and webcams from all around the world

■ News headlines from all around the world

■ Streaming video from a variety of online sources

To install an add-on, navigate to the home page in Raspbmc and hover your mouse over Pictures, Videos, TV Shows, or Music. One of the submenu options is Add-ons; click that.

In the resulting Add-ons window that appears, click Get More.... As shown in Figure 12.16, you'll be presented with an alphabetical assortment of add-on apps that are related to your selected media type.

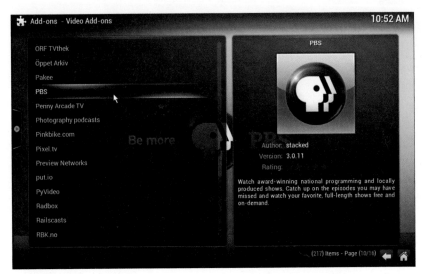

FIGURE 12.16 Add-ons can greatly expand the functionality of your XBMC-based media player system.

Once you've found an add-on that you like, double-click it in the list and then click Install.

To use your new add-on, return to the Add-ons window we saw earlier. Instead of being empty, you should see your new app in the list; double-click it to get started!

## CHAPTER 13

Raspberry Pi Retro Game Station

Let me be completely up-front with you: I am a total and unabashed video game nerd. More specifically, I am a *retro* video game nerd. As much as I enjoy my Xbox, nothing gets my blood pumping quite like firing up my favorite Atari, Coleco, and Mattel Electronics video games from the late 1970s and early 1980s.

As you know from reading this book, the Raspberry Pi itself serves as a physical totem to 1980s microcomputing nostalgia. Therefore, it seems completely natural for us to consider the question of how we can convert the Pi into a retro video game station.

In this chapter I use RetroPie as the software basis for this project. I also show you how to configure the Pi to support joystick controllers.

The RetroPie Project (http://is.gd/kFVq2I) initially started as a plan to turn the Raspberry Pi into a universal retro gaming console that used Nintendo Entertainment System (NES) controllers for input. Although RetroPie supports the emulation of several classic video game consoles, I focus on the Atari 2600 VCS and Nintendo Entertainment System (NES) emulation in this chapter. For the uninitiated, the original console hardware is shown in Figure 13.1.

FIGURE 13.1 At left, the venerable Atari 2600 Video Computer System (VCS), introduced in 1977. At right, the Nintendo Entertainment System (NES), introduced in 1985.

A Word About Console Video Game Emulation

In general terms, *emulation* refers to a computer running one processor architecture to pretend that it is actually a computer running another processor architecture. For instance, the Raspberry Pi port of the RetroArch (http://is.gd/EEb4HQ) multi-system emulator enables the Raspberry Pi, with its ARM processor platform, to play video games from a number of classic video game systems, including

- Atari 2600
- Game Boy Advance
- Intellivision
- MAME
- NeoGeo
- NES
- SNES

Console video game emulation is really quite amazing when you understand that each and every retro video game console had its own proprietary hardware.

The RetroPie Project (http://is.gd/kFVq2I) began as a way to answer the question, "Can we get the Raspberry Pi to play Super Nintendo Entertainment System (SNES) games using the original controllers?"

However, to me the coolest thing about RetroPie is its SD card image, which consists of the following software layers:

- **Raspbian**: This, of course, is the underlying operating system.
- **RetroArch**: This is the console video game emulator. As it happens, the RetroPie image contains a number of other emulators (see http://is.gd/qZrTcQ for a comprehensive list), but I focus on RetroArch in this chapter.
- **Emulation Station** (http://is.gd/a2OmUY): This is the graphical front end for the RetroPie emulator suite; its big advantages are customizability and the ability to control all the menus using only your gamepad (no keyboard or mouse required).

Let's get right into the installation process, shall we?

Installing RetroPie

Here is the high-level overview of the RetroPie installation process:

- Create the RetroPie SD card.
- Boot the Pi and perform initial configuration.
- Populate the ROMs directory. (I teach you more about ROMs later in this chapter, in the section "Transferring ROMs to RetroPie.")

- Customize controls.

- (Optionally) Add media scrapers and other add-ons.

With regard to your Raspberry Pi system requirements, there is nothing too surprising at play:

- Raspberry Pi Model B or Model A (you'll get more performance out of Model B, naturally).

- SD card with at least a 4GB capacity. Some video game ROM files are pretty large, so either invest in a large SD card or plan to store your ROMs on a USB stick plugged into your powered USB hub.

- HDMI connection.

- Keyboard and mouse (for initial configuration; after that, you can control Emulation Station using your joystick or gamepad).

- Gamepad or joystick (more on this later in this chapter, in the section "Setting Up Your Controls").

- Wired or wireless network connection (technically optional but a great convenience nonetheless).

TASK: SETTING UP RETROPIE

In this procedure, you'll download, install, and configure RetroPie on your Raspbery Pi.

1. On your host computer, visit the petRockBlog website and download the RetroPie Project SD card image from http://is.gd/BSyKRP.

Be aware that the SD image is large; it weighs in at approximately 1.4GB. The file comes down as a ZIP file, so you'll need to extract the image file before you can flash it to SD.

2. Use your favorite SD card flashing utility (you did read and study Chapter 4, "Installing and Configuring an Operating System," correct?) to flash the RetroPie image to your SD card.

3. If you have a joystick or gamepad, now is the time to plug it into your Raspberry Pi.

4. Mount the newly flashed SD card into your powered-off Raspberry Pi and boot it up. After seeing the RetroPie splash screen, you are taken automatically into Emulation Station and its text-based wizard to help you configure your joystick or keyboard. I show you what this screen looks like in Figure 13.2.

NOTE: NO JOYSTICK?

If you don't have a joystick, the Emulation Station controller setup wizard will say "No joysticks detected!" In this case, press F4 to quit the controller setup wizard.

5. For each input action you need to perform when using a game, RetroPie asks you to press a key on your keyboard or a button on your joystick (see Figure 13.2).

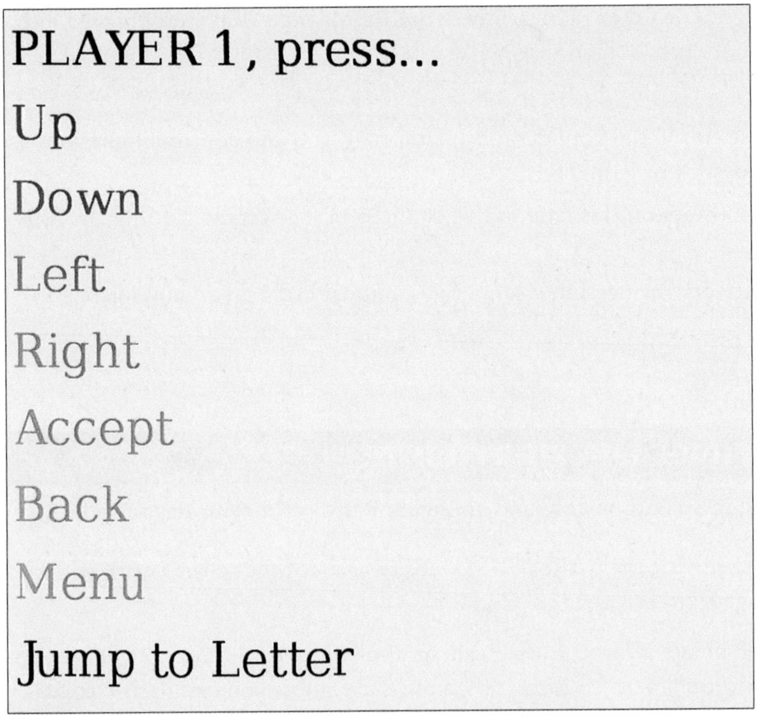

PLAYER 1, press...

Up

Down

Left

Right

Accept

Back

Menu

Jump to Letter

FIGURE 13.2 Emulation Station starts you off with some keyboard or joystick control mapping.

6. You see a "Basic config done!" message when you complete the wizard. You can then press any button or keyboard key to jump into Emulation Station proper.

Don't worry if you feel you made one or more mistakes during the initial controller setup wizard. For instance, your joystick might not have had enough buttons to answer all the setup questions. I show you how to clean up any residual control anomalies a bit later on in this chapter. Breathe easy!

Transferring ROMs to RetroPie

ROMs are the life blood of the retro gaming community. If you ever owned an NES or Atari 2600 or any number of other 80s and 90s game consoles, you know that most of these games came packaged in a small(ish) plastic cartridge.

A ROM image, also called a ROM file, is a bit-for-bit copy of the data from a Read Only Memory (ROM, get it?) chip from these cartridges. According to copyright law, to legally download a ROM, you must already possess the original game cartridge for any ROM file that you have in your possession for play with an emulator. Fortunately, there's good news if your mom threw out your old video game collection when you left home: There are any number of outlets, both online and brick and mortar, that deal in used and vintage game cartridges.

NOTE: HOW DO I CREATE A ROM?

Okay, let's assume you took down that boxful of ancient Atari 2600 cartridges from a dusty box in your garage attic. What now? How can you take those legally owned games and convert them to ROM files that are playable in RetroPie?

I'm glad you asked! The general workflow is that you need to find a way to read the ROM from your source cartridge (obviously, right?) and then copy that tiny amount of data either to a blank cartridge or directly to your connected computer. This process always involves dedicated hardware, and sometimes requires you to download schematics and assemble PCBs yourself. Personally, I feel ethically safe by downloading the ROMs from a known source like AtariAge.com so long as I also own the corresponding game cartridges.

For now, I assume that you have one or more game ROMs that you'd like to load on your RetroPie device. How do you do that?

Well first of all, remember that RetroPie uses Raspbian under the hood as the host operating system, so your first order of business is to run sudo raspi-config and perform the initial setup of the device. If you need a refresher, just take a look back at the raspi-config sections in Chapter 4.

Next, you should update the system software and reboot:

```
sudo apt-get update && sudo apt-get upgrade
sudo reboot
```

NOTE: ENHANCING PERFORMANCE

You might also want to consider adjusting the Pi's memory split to favor graphics over processor performance, as well as enabling overclocking. I cover both of these subjects in exhaustive detail in Chapter 18, "Raspberry Pi Overclocking."

As you know, Raspbian enables Secure Shell (SSH) connections by default. Thus, you can use a Secure Copy (SCP) utility to transfer ROM files to the Pi.

TASK: TRANSFERRING ROMS TO YOUR RASPBERRY PI

1. On your Raspberry Pi, if you are in Emulation Station, press F4 to exit to a Terminal prompt. Next, type **ifconfig eth0** to obtain the Pi's IP address.

2. On a remote computer that hosts your ROM files, use FileZilla, Cyberduck (http://is.gd/ JDzRgw), or another SFTP client tool to connect to your Pi. Be sure to specify SFTP, and not FTP, as the connection method.

   The default username and password are the ones you would expect: username, pi; password, raspberry.

   Take a look at Figure 13.3 to see how I set up my connection in FileZilla. To create a new stored connection, open FileZilla and click File, Site Manager, New Site.

FIGURE 13.3 FileZilla enables you to store connection information permanently for convenience.

3. Navigate to the path /home/pi/RetroPie/roms. You see that RetroPie creates folders to store ROMs for all its supported console video game platforms.

4. You can now drag and drop your ROM files into the appropriate subfolders. Note that you need to upload the actual binary ROM files to the Pi. For instance, NES ROMs typically have the .nes file extension, but Atari 2600 ROMs usually employ the .bin file extension. You can see my system in Figure 13.4.

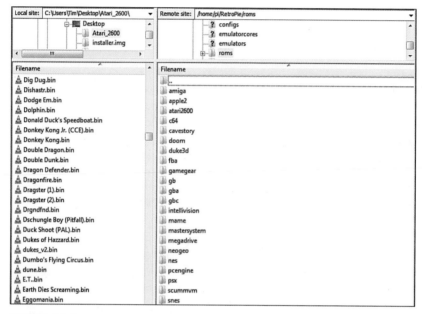

FIGURE 13.4 Here you can see where ROMs should be stored on your RetroPie machine.

5. You can start Emulation Station from the Terminal prompt of your Pi by issuing the following simple command:

```
emulationstation
```

Remember that Linux is case-sensitive; thus EmulationStation or EMULATIONSTATION generate errors, but not Emulation Station.

By default, Emulation Station/RetroPie includes the following games:

- Cave Story
- Doom
- Duke Nukem 3D (demo)

You can also start the LinApple Apple II emulator (http://is.gd/Ea0UUT) and the rpix86 DOS 5.0 emulator (http://is.gd/1vGqRY) and run old Apple II games.

You can use your mapped LEFT and RIGHT controls to switch among the installed games and use your mapped ACCEPT button to start one. More on gameplay later, though—one step at a time!

If you run into any expected results in the ROM detection process, your first step should be to open up the ~/.emulationstation/es_systems.cfg configuration file for editing.

The two parameters you want to watch for are

- **PATH**: This is the default location where Emulation Station expects to find game ROMs for each platform. If this is set incorrectly, you won't be able to see your games.

- **EXTENSION**: You need to ensure that your game ROMs all have one of the supported file extensions for detection to complete properly. For instance, the most common Atari 2600 ROM file extensions are .bin, .img, and .z26. NES file extensions are typically, reasonably enough, .nes.

NOTE: SHOWING FILE EXTENSIONS IN WINDOWS

In Windows 7 or Windows 8, file extensions are hidden by default. To show them, open the Folder Options Control Panel and navigate to the View tab. Under Advanced Settings:, enable the option Hide extensions for known file types and click OK to confirm the change.

In OS X, choose Finder > Preferences, and then navigate to the Advanced pane. Next, enable the option Show all filename extensions.

If you make any changes to the file, don't forget to save your changes and reboot your Pi before attempting another scrape.

Before I get to playing games (I know you are as excited to do that as I am), let's revisit how to tweak up the keyboard and, more importantly, the joystick controls.

Setting Up Your Controls

Emulation Station is optimized for joystick/gamepad-based control. However, for the sake of completeness I want to show you how to edit the keyboard mappings.

From Emulation Station, press F4 to exit to a Terminal prompt. The RetroPie controls (all of them, keyboard and joystick) are stored in a configuration file named retroarch.cfg. Use the following command to edit the file:

```
sudo nano ~/RetroPie/configs/all/retroarch.cfg
```

Of course, you can also use your SFTP utility to download a copy of the file to your remote computer, edit the file using your favorite text editor, and reupload the file to the Pi, overwriting the old version.

In any event, look for the line that starts with # **Keyboard input**. You can edit the key mapping values directly here; look at Figure 13.5 to see my setup.

FIGURE 13.5 The retroarch.cfg file is where all RetroPie control defaults are stored.

NOTE: EXTRA, EXTRA, READ ALL ABOUT IT!

You should study the full contents of the retroarch.cfg file because you can actually make some pretty cool changes to RetroPie. For instance, visit the #Saves state section to customize key or joystick mappings to save and load game state. Very useful!

Now let's turn our attention to joystick mappings.

TASK: CONFIGURE RETROPIE JOYSTICK CONTROL MAPPINGS

You can use the retroarch-joyconfig utility to customize joystick mappings. Let's do that now.

1. From the Terminal prompt on your RetroPie computer, navigate to the appropriate directory location.

```
cd ~/RetroPie/emulators/RetroArch/tools
```

2. If you run retroarch-joyconfig with no parameters, the results of your configuration are dumped to the screen but are not saved in the retroarch.cfg file. That isn't cool. Thus, you need to redirect the output of the retroarch-joyconfig program directly to the retroarch.cfg file like so:

```
./retroarch-joyconfig >> ~/RetroPie/configs/all/retroarch.cfg
```

3. You are prompted to assign bindings for each command; do so by pressing the appropriate button on your joystick. A screen capture (not the greatest) of my monitor is shown in Figure 13.6.

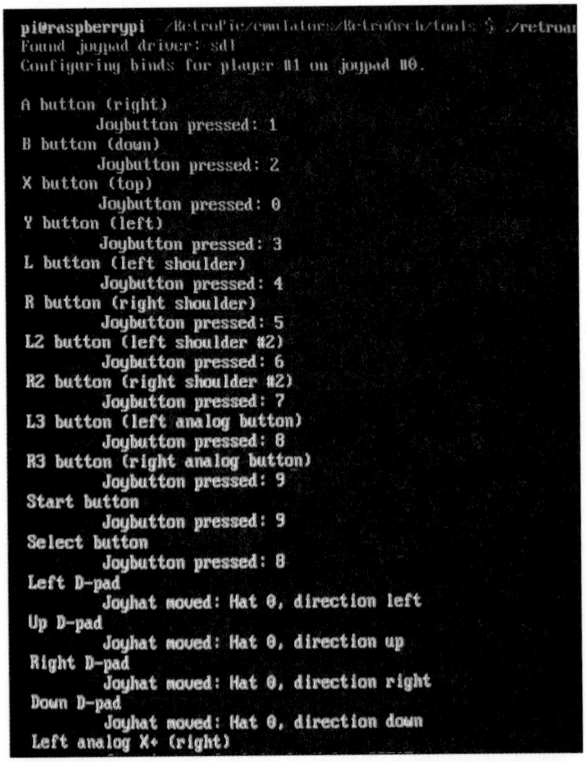

FIGURE 13.6 Establishing your joystick key bindings

You can always edit the retroarch.cfg file afterward to remove or comment out lines that aren't relevant to your joystick. To comment out (and therefore nullify without deleting) an entry, simply prepend the line with an octothorpe (#) character.

There's one more edit you should consider making to retroarch.cfg because you should definitely add a joystick mapping that allows you to exit your active emulator and return to Emulation Station. After all, you shouldn't have to reboot the Pi to revisit your game menus.

NOTE: DON'T FORGET STARTX

If you don't like using nano or another text-based text editor, you can always type **startx** and use the GUI tools in LXDE to accomplish your RetroPie configuration. Just remember to dump X and return to the Terminal shell when you're finished. (As a reminder for doing so, you can click the red Power button in the lower right corner of LXPanel.)

Go to the end of the retroarch.cfg file and add the following two lines:

```
input_enable_hotkey_btn = "X"
input_exit_emulator_btn="Y"
```

Substitute X and Y for two joystick buttons that you'll press simultaneously to exit the emulator and return to Emulation Station. (And don't forget what you selected!)

Playing Your Games

If you haven't already done so, reboot your Raspberry Pi or type **emulationstation** from the Terminal prompt to start Emulation Station. Here's the deal:

- **Use the LEFT and RIGHT controls to scroll through the emulator menus**: You only see an entry for emulators that actually contain ROMs. Thus, on my system I have game lists for Atari 2600 and NES because I uploaded ROMs for those platforms.

- **Use the UP and DOWN controls to scroll through the game menus, and use the ACCEPT and SELECT controls to launch and start games**: If you mapped the PAGE UP, PAGE DOWN, or START WITH LETTER mappings, it makes it easier to locate games in huge ROM lists.

- **Use ESC or whatever custom joystick mapping you specified to exit the emulator and return to Emulation Station**: This is a particularly important option because you should be able to return to Emulation Station without having to reboot the Raspberry Pi.

Figure 13.7 shows you a typical game screen.

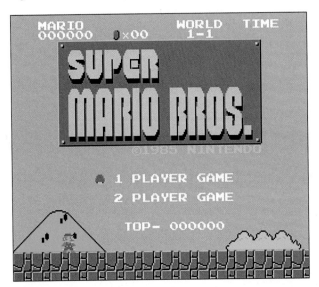

FIGURE 13.7   This is Super Mario Brothers, one of the most popular video games of all time.

I've found that the performance of the retro video games is (at least) as good as it is on original hardware. Does that surprise you? Think of it this way: Even the Model A Raspberry Pi board is orders of magnitude more powerful than, say, the Atari 2600 or the NES (SNES, for that matter).

One thing—you might be put off by the slightly warped display of old 8-bit games on your widescreen monitor. By default, the games fill the entire screen. To tweak up the emulator resolution, open retroarch.cfg for editing, and check out the #### Video section.

Notice that most of the configuration entries are commented out (that is to say, deactivated). To activate an option, simply remove the octothorpe (#) preceding the appropriate line.

Installing Useful Add-Ons

In Chapter 12, "Raspberry Pi Media Center," you learned how useful media scrapers are to fill in the blanks on media content. Did you know that you can use media scrapers with your retro video game ROMs as well?

Yes, indeed. Not only can you play your favorite old-school games, but you can see the original box art, learn trivia about the game's history, and much more.

To truly dig into all possible retro video game goodness, let me show you how to install the ES-scraper utility to scrape your ROM directories and download box art and game descriptions.

TASK: INSTALL ES-SCRAPER

1. From Terminal on your Raspberry Pi, navigate to the appropriate directory:

```
cd (to ensure you're in your home directory)
cd RetroPie/supplementary
```

2. Create a local copy of the ES-scraper online repository and run the RetroPie setup Linux shell script:

```
git clone http://github.com/elpender/ES-scraper
cd
cd RetroPie-Setup
sudo ./retropie_setup.sh
```

3. When you're in the RetroPie Setup utility, use the Tab, number, and Enter keys to navigate the text menus.

In the Choose installation either based on binaries or on sources dialog, select Setup (only if you already have run one of these installations).

4. In the Choose task dialog, select Run 'ES-scraper' as shown in Figure 13.8.

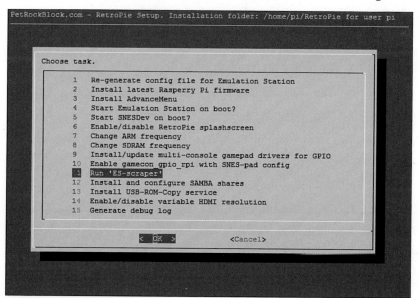

FIGURE 13.8   You can force a game ROM metadata scrape from within the RetroPie Setup script.

5. Select (Re-)scrape of the ROMs directory to perform an immediate ROM discovery and metadata download.

The time required for ES-Scraper to complete its metadata scrape and resource download depends on the number and type of game ROMs you have available on your Raspberry Pi.

When the process is complete, you can cancel out of the RetroPie Setup script, reboot your Pi, and enjoy the new artwork! An example of downloaded game descriptions and box art is shown in Figure 13.9.

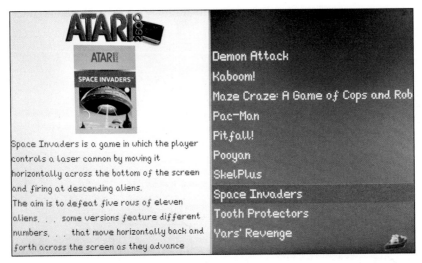

FIGURE 13.9 The downloaded box art and game description data makes browsing your ROM collection that much more enjoyable.

In Search of the Perfect Joystick

In my experience, RetroPie does an excellent job of detecting your USB joystick or gamepad. I have used many, many game controllers over the years, and for my money nothing beats my old Logitech Rumblepad 2, which employs the classic Playstation/PS2 form factor.

NOTE: IN CASE YOU WONDERED...

If you've wondered throughout this chapter, "What's the difference between a gamepad and a joystick?" let me clear up any residual confusion. A gamepad, also called a joypad, is a game controller that is typically operated with two hands. Gamepads usually have an 8-way digital pad (d-pad) as well as one or two analog sticks. By contrast, a joystick consists of a single 8-way or analog control handle, with or without additional action buttons or triggers.

On the other hand, many retro video game purists want to enjoy their favorite emulated games by using either a reproduction or original controller from the original consoles.

Quality varies widely for the USB reproductions; most of us prefer adapters that transform the proprietary controller plugs into USB. In Figure 13.10 you can see a mashup of some of my favorite video game controllers.

FIGURE 13.10   Some of my favorite video game controllers: (1) Logitech Rumblepad 2; (2) Atari 2600; (3) SNES; (4) Sega Genesis.

RetroZone (http://is.gd/Cs2GKf) sells USB adapters for the following console video game controllers:

- Atari 2600
- NES
- Nintendo 64
- Sega Genesis
- SNES

Please note that you still need to purchase the original controller in addition to buying an adapter. They don't all rate so well among hardcore classic gamers, but some companies produce reproductions of old controllers with native USB connectivity:

- Tomee NES USB Controller (http://is.gd/DkrM9c)
- Tomee SNES USB Controller (http://is.gd/kEZg3P)

- Retrolink NES USB Controller(http://is.gd/1kuzKi)

- Retrolink SNES USB Controller (http://is.gd/PJUB0h)

- Retrolink Nintendo 64 USB Controller (http://is.gd/68XUFP)

- Retrolink "Classic Controller" (modeled on Sega Genesis controller) (http://is.gd/3HzHns)

The people behind the Retropie Project have developed a GPIO adapter for the original Super Nintendo Entertainment Center (SNES) controllers. You can get all of the details and assembly instructions at their website at http://is.gd/clRqqZ.

As you can see (at least in part) in Figure 13.11, the unit consists of an adapter PCB, two SNES connectors, a couple ribbon cables, and a ribbon crimp connector.

FIGURE 13.11 The PetRockBlog RetroPie GPIO Adapter

The GPIO adapter includes an extra tactile pushbutton that can be useful, for instance, to map to the EXIT EMULATOR command, which of course allows you to cleanly close the emulator.

From the looks of their documentation, the mapping between the SNES controller pinout and the Raspberry Pi pinout is pretty straightforward. Check out Figure 13.12 to judge for yourself. Basically you are soldering each SNES connector lead to a ribbon cable, which in turn connects to particular pins on the Raspberry Pi GPIO header.

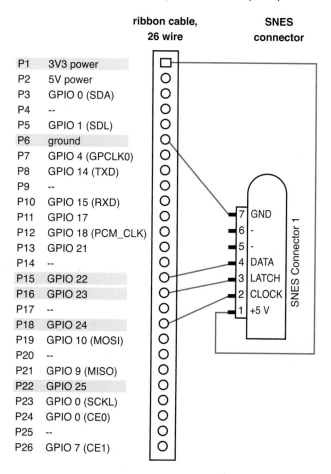

FIGURE 13.12 RetroPie GPIO Adapter pinout schematic

If you can catch them while they have units in stock, the PetRockBlog also sells fully assembled units for $18.40 USD. All you have to do with the purchased units is to solder the SNES controller connectors to the included ribbon cable.

## CHAPTER 14

Raspberry Pi *Minecraft* Server

Minecraft, a sandbox construction game originally created by the Swedish programmer Markkus "Notch" Persson (http://is.gd/Y7W6Gy) and later absorbed by Notch's company Mojang AB (http://is.gd/07x9Au), is more than a game: It is a phenomenon. In a gaming context, "sandbox" means that the game has no defined storyline; instead, players can roam around the game world and do pretty much whatever they want (see Figure 14.1).

As a game developer you know you are onto something when your product is used in school as well as home, and gamers collaborate with each other and lose sleep playing it.

FIGURE 14.1 *Minecraft* 1.5.2 for Microsoft Windows

Retro video game nerds such as myself deeply appreciate the chunky 8-bit graphics. In fact, the visual presentation of *Minecraft* reminds me of a cross between *Super Mario Brothers* for the Nintendo Entertainment System and *Doom* for the PC.

As the game title suggests, the two main tasks in *Minecraft* are mining, which involves breaking various and sundry ore blocks in search of useful raw materials, and crafting, which means taking mined raw materials and fashioning tools, weapons, furniture, food...you name it.

Minecraft includes two primary game modes:

- **Survival**: The player's avatar can die, and the world is inhabited by enemy nonplayer characters (NPCs) called *mobs* that can destroy the player. The player also starts with no tools or raw materials but can install other user-made modifications (mods) to change how the game world works. The focus on this mode is on exploration, combat, resource gathering, and construction.

- **Creative**: The player's avatar cannot die and is capable of flight. No mobs exist in this world. The player's inventory includes all items found in the game, including materials, eggs, potions, and so forth. Because Creative mode involves no combat and has no need to forage for tools and resources, the focus here is on creative construction.

Minecraft includes two additional gameplay modes, Adventure and Hardcore, that aren't completely fleshed out as of this writing in summer 2013. You can read more about *Minecraft* game modes by visiting the *Minecraft* Wiki at http://is.gd/8kEFWj.

NOTE: LEARN HOW TO PLAY *MINECRAFT*

In this chapter, I provide only the most cursory of introductions to *Minecraft* and assume you have at least a passing familiarity with its gameplay. For a complete introduction to the game, please read my Pearson colleague Stephen O'Brien's excellent book (which I tech-edited it, in fact), *The Ultimate Player's Guide to Minecraft* (http://is.gd/yvXXbl).

In general, I think the main reasons why *Minecraft* is so popular are the following:

- Players can assert and flex their creativity.
- The game enables players to create objects and share them with other players around the world.
- The game is extensible, allowing proficient players to broaden and deepen the game world.

Minecraft is also used in primary, secondary, and higher education. Why? Let's count some of the ways:

- The game teaches problem-solving skills in a manner that is engaging and fun.
- It teaches players how to use code to modify the behavior of a system (*modding*; more on that subject later).
- The game itself can be used as an instructional tool. For instance, a teacher can build lessons inside a shared *Minecraft* game world, and the students can interact with the lesson as avatars.

The Mojang business model for *Minecraft* is to give players what they want and port the game to as many different computing platforms as possible. Check it out:

- *Minecraft*: The original game is coded in Java and runs on Windows, OS X, and Linux.

- *Minecraft*-**Pocket Edition**: This is a heavily scaled-back edition of *Minecraft* that is coded in C++ and runs on iOS (iPhone, iPod touch, iPad) and Android.

- *Minecraft*: **Xbox 360 Edition**: This is not only the full version of *Minecraft*, but the game also includes several features that are specific to the Xbox 360 port, such as simpler crafting mechanics, in-game tutorials, and robust split-screen and Internet multiplayer gaming.

- *Minecraft*: **Pi Edition**: This is an educational *Minecraft* port that is largely unlocked and allows the gamer much greater control over the game world compared to the other editions of the game.

Naturally, this book is all about the Raspberry Pi, so I am constraining the discussion in the remainder of this chapter to *Minecraft*: Pi Edition.

The *Minecraft* server forms the basis of the multiplayer aspect of the game. Therefore, I also cover how to build a Raspberry Pi-powered *Minecraft* game server.

Let's get to work!

Installing *Minecraft* Pi

According to the documentation at the *Minecraft* Pi website (http://is.gd/ORylMx), *Minecraft* Pi has been optimized to run under the official Raspbian Linux distribution. Therefore, if you haven't already flashed your SD card and gotten your Raspbian-based Pi up and running, please do that first before proceeding. Remember that we learned how to flash SD cards in Chapter 4, "Installing and Configuring an Operating System."

Before you begin, it's important to note that *Minecraft* Pi doesn't work over a VNC connection, so make sure you boot your Pi with a monitor, keyboard, and mouse attached.

TASK: INSTALLING *MINECRAFT* PI

In this procedure you'll get *Minecraft* Pi up and running on your Raspberry Pi. For obvious reasons, you'll be working from an LXDE graphical shell here.

1. If you aren't already in LXDE, type **startx** from the Bash shell prompt to get into GUI mode.

2. Make sure you are in your home directory, and then make a directory for the game:

```
cd
mkdir Minecraft
cd Minecraft
```

3. Fire up LXTerminal and download the software using the nifty wget utility:

```
wget -O Minecraft-pi-0.1.1.tar.gz [ic:ccc]https://s3.amazonaws.com/assets.
Minecraft.net/pi/[ic:ccc]Minecraft-pi-0.1.1.tar.gz '
```

NOTE: A MATTER OF FORMATTING

When I give you single command statements such as the wget statement in step 3, don't use the Enter or Return key until you've typed in the entire statement. In other words, ignore the line breaks in the book unless specifically instructed not to.

---

4. Use tar to extract the contents of the tar.gz archive you just downloaded:

```
tar -zxvf Minecraft-pi-0.1.1.tar.gz
```

5. Delete the tar.gz archive, navigate into the newly extracted game directory, and start the game!

```
rm Minecraft-pi-0.1.1.tar.gz
cd mcpi
./Minecraft-pi
```

Notice that in Linux, you run executable programs from the current working directory using the dot slash notation. This shorthand notation enables you to run programs without (a) having to supply the entire path to the executable, or (b) having to put the app path in the system's PATH environment variable.

Minecraft's Home menu screen is displayed in Figure 14.2.

FIGURE 14.2 *Minecraft* Pi Edition

To play, click Start Game and click Create New to build a new world. Here are your basic controls (you can also read ~/Minecraft/mcpi/CONTROLS.txt):

- **Mouse**: Turn your avatar.
- **Left Mouse Button**: Remove block.
- **Right Mouse Button**: Place/hit block.
- **Mouse Wheel**: Select inventory item.
- **W, A, S, D**: Move your avatar forward, backward, left and right, respectively.
- **SPACE**: Jump.
- **E**: Open inventory.
- **1-8**: Select inventory slot to use.
- **ESC**: Show or hide game menu.
- **TAB**: Release mouse.

Minecraft Pi Edition is built from the *Minecraft* Pocket Edition code base, so if you played *Minecraft* on your iOS or Android device, then you pretty much understand how the game works on the Raspberry Pi. One significant limitation of the *Minecraft* Pi Edition, at least in the initial 0.1.1 release, is that the game supports only the Creative game mode.

Accessing the Python API

An *application programming interface* (API) is a set of rules that define how a user can access and potentially modify the default code base for an application. As I said earlier, *Minecraft* Pi Edition was developed as a way to teach people how to learn computer programming in the context of game development. Accordingly, the good people at Mojang include class libraries for both Python (located in ~/Minecraft/mcpi/api/python/mcpi) and Java (located in ~/Minecraft/mcpi/api/java) programming languages.

NOTE: HEAD OF THE CLASS

In object-oriented programming (OOP) terminology, you can look at a *class* as a template that describes the attributes (properties) and behaviors (methods) of an object. All objects in *Minecraft* are originally defined as classes. A *class library* is simply a code file that is filled with class (object) definitions.

The Python class libraries are Python 2, not Python 3, but don't worry about that; everything you learned in Chapters 10, "Programming Raspberry Pi with Python—Beginnings," and 11, "Programming Raspberry Pi with Python—Next Steps," still applies. I'm just speculating, but I believe that Mojang chose Python 2 over Python 3 because they wanted the class libraries to reach the widest possible audience. After all, most OS X and Linux distributions include Python 2 by default.

In short, you have a handful of Python and Java scripts that provide the *Minecraft* player with tools to control the game world. This is pretty cool stuff, so let's dive right in, shall we?

TASK: LOADING THE *MINECRAFT* PYTHON LIBRARIES

Here we will make a copy of the Python class libraries and create a simple script that pops a chat session into an active game. The examples in this section are adapted from Martin O'Hanlon's wonderful work at his Stuff about Code blog (http://is.gd/Y2nUFZ).

1. Run *Minecraft* Pi Edition and start a new game. You must be in world to see any results of your API programming.

2. Press the TAB key to escape the game and free you up to go elsewhere in LXDE.

3. Open an LX Terminal session and create a working directory for your scripts and copy the API files into the new folder:

```
cd
mkdir Minecraft-magpi
cp -r Minecraft/mcpi/api/python/mcpi/ Minecraft-magpi/Minecraft
```

I'm assuming that *Minecraft* Pi exists in the path ~/*Minecraft*. Also note that the API library files need to be stored in a subdirectory called *Minecraft*.

4. Let's create a new script file:

```
sudo nano Minecraft-magpi/mctest.py
```

5. Populate the file like so:

```
#!usr/bin/env python
import Minecraft.Minecraft as Minecraft
import Minecraft.block as block
import time

mc = Minecraft.Minecraft.create()

mc.postToChat("Hello, World of Minecraft!")
time.sleep(5)
```

Whew—that is a lot of code. Let's take it line by line:

1: This is the traditional "shebang" line that gives the operating system direction for finding the Python interpreter

2–4: Import relevant modules. The first two calls pull two classes from the API libraries; one for the *Minecraft* world itself, and the second one for the *Minecraft* block. The time module is built into the Python default class libraries.

5: Instantiate (or bring into being) an instance of the *Minecraft* world, packed into a variable named mc. This code essentially connects you to the running *Minecraft* instance on the Pi.

6: Use the postToChat method of the *Minecraft* object to send a chat message to the game session.

7: The sleep function controls how long you want your chat message to stay on screen.

Save your work and close the script file when you're finished.

6. Now let's test by running the script. Make sure to switch focus back to the game screen to get the full effect.

```
python Minecraft-magpi/mctest.py
```

You can view the output in Figure 14.3.

FIGURE 14.3   Using Python to interact with the *Minecraft* world

The overall *Minecraft* Pi Edition Python API specs can be found in the file ~/Minecraft/mcpi/ api/spec/mcpi_protocol_spec.txt. This file explains how all of the API functions work. It's recommended reading, for sure.

How about another example? Note that the player's avatar coordinates are displayed in the upper-left corner of the screen. Check out Figure 14.4 for a visual explanation of what these coordinate values mean.

pos: 2.8, 2.0, -1.8

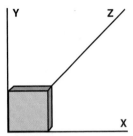

FIGURE 14.4 *Minecraft* Pi Edition displays the player's location onscreen by using x, y, and z coordinates.

As you can see by studying Figure 14.4, coordinates denote an object's specific location within the *Minecraft* world.

TASK: MAKING YOUR PLAYER JUMP HIGH!

In this task you'll play with the *Minecraft* environment by modifying *Minecraft* such that your avatar is thrown high in the air like a cannonball.

1. Reopen the mctest.py script file you created in the previous exercise and open it up in nano or your favorite text editor.

2. Make the code in your script file look like mine:

```
#!usr/bin/env python
import Minecraft.Minecraft as Minecraft
import Minecraft.block as block
import time

playerPos = mc.player.getPos()
mc.player.setPos(playerPos.x, playerPos.y + 100, playerPos.z)
mc.postToChat("You are gonna fall!")
time.sleep(5)
```

There are two new lines of code here (lines 5 and 6) relative to the previous task:

- Define a variable that stores the player's current onscreen position (x, y, and z coordinates)

- Adjust the player's position 100 blocks along the y (vertical) axis. This has the effect of boosting the avatar high into the air.

3. When you run the script, be sure to take control of the game within five seconds because your avatar is going to typify the old law "What goes up must come down!"

How about we do one more quick example, this one demonstrating how you can alter the mining and crafting aspects of the game.

TASK: CREATE A DIAMOND FLOOR

1. Again, open up your previous mctest.py script file, this time editing the contents to match the following:

```
#!usr/bin/env python
import Minecraft.Minecraft as Minecraft
import Minecraft.block as block
import time

PlayerPos = mc.player.getPos()

PlayerTilePos = mc.player.getTilePos()

mc.setBlocks(playerTilePos.x - 25, playerTilePos.y - 1, playerTilePos.z - 25,
playerTilePos.x + 25, playerTilePos.y -1, playerTilePos.z + 25, block.DIAMOND_
BLOCK)
mc.postToChat("Now thats a big diamond floor!")
```

2. Switch to your game screen to verify that the code has gone into effect. The stunning, valuable result is shown in Figure 14.5.

FIGURE 14.5   The world is your oyster...or diamond...in *Minecraft* Pi Edition.

You learned in the previous example that mc.player.getPos() determines the current coordinates of the player.

The PlayerTilePos variable determines which tile the player is currently standing on.

The setBlocks function is pretty robust; be sure to read the documentation to get the names of all the block types. The generic formulation of the function is

```
setBlocks(x1, y1, z1, x2, y2, z2, blockType, blockData),
```

This code takes two sets of coordinates and fills the gap between them with a particular block type. In this case it creates 25 diamond blocks in front of, behind, to the left, and to the right of the player, which places the player directly in the center of a big, diamond square. Pretty cool, eh?

Building a *Minecraft* Server

At its core, *Minecraft* is a single-player game. However, multiplayer functionality is built into the platform, which enables more than one player to coexist in the same game world. What's different about *Minecraft* multiplayer from, say, *Call of Duty*, is that in *Minecraft* players tend to work cooperatively in building things rather than against each other.

Numerous public *Minecraft* servers are available for connection, and you can always download the free *Minecraft* Multiplayer Server software for Windows, OS X, or Linux at http://is.gd/Scuod8.

The challenge to making your Raspberry Pi a *Minecraft* server is, naturally, the board's hardware resource limitations, but also the heaviness of Java. As it happens, the multiplayer server software is a Java server application, so you need to do some extra homework to make this happen using your battle-weary Pi.

To prepare your Pi for duty as a *Minecraft* server, make the following tweaks to your system:

- If you can, use your Ethernet cable to connect to your local area network instead of a Wi-Fi dongle. You'll get more reliable data transmission and speed.

- Run sudo raspi-config and adjust the CPU/GPU memory split in favor of the CPU. Some server operators suggest setting the GPU to only 16MB. Then overclock the Pi as much as you dare.

  I cover the CPU/GPU split and overclocking in detail in Chapter 18, "Raspberry Pi Overclocking."

- Make sure your system is current by running

  ```
  sudo apt-get update && sudo apt-get upgrade
  ```

- If you've installed applications and services that run in the background, consider reflashing that SD card or loading up another card with a pristine install of Raspbian. You don't need unwanted cruft slowing down your *Minecraft* server.

You have the decision whether to install a stock *Minecraft* server using the installer provided by Mojang or to install a modified version. The two most popular *Minecraft* server alternatives are

- **CraftBukkit** (http://is.gd/Dpm6VE): Modified version of the Mojang *Minecraft* server file; allows for plugins and various other extensions to the *Minecraft* server environment. Specifically, CraftBukkit is the Bukkit server executable, and Bukkit represents the programming API.

- **Spigot** (http://is.gd/Nj654R): Modified version of the Bukkit API; optimized for smaller servers (like the Raspberry Pi!).

In this chapter I take you down the Spigot route. That won't exempt you from the Java requirement, but you'll have a much leaner, cleaner, meaner, and better performing *Minecraft* server.

TASK: INSTALLING JAVA AND THE MINECRAFT SERVER

1. From a shell prompt, verify that you don't have Java installed:

```
java -version
```

This command should throw an error if Java is not present on the system.

2. Pull down and install an appropriate Java distribution now, after first ensuring that your Pi has the appropriate certification authority (CA) certificates:

```
sudo apt-get install ca-certificates
sudo wget http://www.java.net/download/JavaFXarm/jdk-8-ea-b36e-linux-arm-hflt-29_
nov_2012.tar.gz
```

This installs a version of Java that Oracle developed expressly for the Raspberry Pi; read more about it at the Java.net website: http://is.gd/L8T7fJ.

3. That Java package name is huge, so rename it to make it more manageable:

```
mv jdk-8-ea-b36e-linux-arm-hflt-29_nov_2012.tar.gz jdk.tar.gz
```

NOTE: TAB COMPLETION ROCKS

I know that I've mentioned this before, but it bears repeating: You can double, triple, or quadruple your Linux command-line navigation if you get into the habit of pressing Tab after typing the first few characters of a folder or file name. Tab completion works; it really does!

4. Perform some housekeeping and actually install Java:

```
mkdir -p /opt
sudo tar zxvf jdk.tar.gz /opt
rm jdk.tar.gz
```

5. Verify you have Java installed:

```
sudo /opt/jdk1.8.0/bin/java -version
```

6. Cool! With Java installed, you're halfway home. Let's now install Spigot:

```
cd
sudo wget http://ci.md-5.net/job/Spigot/lastBuild/artifact/Spigot-Server/target/
spigot.jar
```

By the time you read this, md-5 will have released a newer build of Spigot. Thus, keep the website http://is.gd/3nF2tr bookmarked and edit the URL just given to reference the latest and greatest build.

7. It's time to start the *Minecraft* server. Doing so creates the server.properties file from which you can tweak the server's behavior.

```
sudo /opt/jdk1.8.0/bin/java -Xms128M -Xmx256M -jar /home/pi/spigot.jar nogui
```

In this statement, you start the Java virtual machine using a RAM footprint of between 128MB on the low end and 256MB on the high end. The nogui parameter is important because, of course, you need to run your server as lean and mean as possible to conserve system resources.

Expect it to take several minutes for the server to fully generate the *Minecraft* environment. While the server bootstraps, you see hundreds of lines of output scroll in your Terminal window. Do not be alarmed.

8. To test the server, start *Minecraft* from a remote system, click Multiplayer in the splash screen (Join Game in *Minecraft* Pi Edition), and select your Raspberry Pi server from the server list (see Figure 14.6).

FIGURE 14.6 You can connect to our multiplayer server either locally or remotely.

NOTE: VERSION CONTROL AGAIN, AND YET AGAIN

If you see the "Server Outdated" error message when you try to connect to the server from the *Minecraft* client, you should download the latest version of the Spigot software. You can keep abreast of version releases at the Spigot website at http://is.gd/3nF2tr.

9. Double-click the Pi server in the list, and you're logged in!

Note that if you want to advertise your *Minecraft* Pi Server to the Internet, you need to configure your router to forward traffic on Transmission Control Protocol (TCP) port 25565. You can learn how to configure an internal network device with a public IP address by reading Chapter 15, "Raspberry Pi Web Server." You can learn how to configure port forwarding on your router by visiting PortForward.com (http://is.gd/ttSr5H).

Administering a *Minecraft* Server

Minecraft server administration is an art and science unto itself and is therefore far outside the scope of this book. Nevertheless, I want to give you the core need-to-know information.

In the Terminal window from which you started the *Minecraft* server, type **help** to get a list of *Minecraft* server commands.

A *Minecraft* server operator is known as an op (pronounced *op* or *oh-pee*, and sometimes stylized as OP). Before you start issuing online commands, however, you should learn how to modify the *Minecraft* server configuration file.

Type **stop** in the *Minecraft* Server console to stop the server. Next, open the config file, which is located by default in your home directory:

```
sudo nano server.properties
```

The server.properties file consists of simple key/value pairs; the trick is learning what each property means. I suggest you review the list at the *Minecraft* Wiki (http://is.gd/awZBsZ).

Visually, the file isn't much to look at; it's just a typical plain text configuration file. For instance, here are the first few lines of a sample server.properties file:

```
#Minecraft server properties
#Wed May 22 21:15:19 EDT 2013
generator-settings=
allow-nether=true
level-name=world
enable-query=false
```

NOTE: OPERATING A *MINECRAFT* SERVER

In *Minecraft* server nomenclature, an operator, or OP (oh-PEE, or ohp) is a superuser who has full control over the entire server. Obviously, you as the server owner should have OP privileges, but you should be very careful before assigning OP to any other *Minecraft* users.

Make your *Minecraft* user an OP by typing the following command in the server console:

```
op <username>
```

You see this feedback display onscreen directly in your game session. To issue a server command in the game, precede the command with a slash (/). For instance

```
/me <message>
```

This command sends a status message to all connected players on the server. You can use /tell to send private messages to individual users. The in-game multiplayer experience is represented in Figure 14.7.

FIGURE 14.7   You can issue player or op commands directly in the game.

If you find the *Minecraft* server command syntax similar to that of Internet Relay Chat (IRC), then good for you—that's exactly what it feels like.

As an OP, you have godlike control over the server-spawned Minecraft world. For instance, if it's nighttime and you want to jump time to dawn, try this from the game:

```
/time set 0
```

A time value of 12000 takes you to dusk.

MinecraftServerHost.net provides a good, comprehensive list of *Minecraft* player and op server commands at http://is.gd/ttSr5H. The trusty *Minecraft* Wiki is also helpful (http://is.gd/ax3Lrr).

Minecraft Plugins

Finally, we come to the subject of plugins, which are add-ons to the *Minecraft* server that enable you to vastly extend your control over the multiplayer gaming environment. Many *Minecraft* OPs search for plugins by browsing the Bukkit website at http://is.gd/P6l0Rs. Here is the high-level installation overview:

- Download the plug-in .JAR file to your server.
- Place the .JAR file in your plugins directory.
- Stop and start the server.

## CHAPTER 15

Raspberry Pi Web Server

A *web server* is a computer that serves content by using standard Internet protocols. The word standard is key here because web communications are shared seamlessly across any kind of device, from desktop computers to video game consoles to tablets and mobile phones.

Any device that can (a) connect to a local area network or the Internet, and (b) has a web browser or web-aware application installed makes use of these standard protocols. The content that is served by a web server consists of the following media types at the very least:

- **Web pages**: Text that is formatted with hyperlinks, pointers to other content on the same page, the same website, or a different website (representative file types: .htm, .html, .php, .aspx).

- **Images**: These can be static or animated pictures of the bitmap or vector variety (representative file types: .gif, .jpg, .png, .svg).

- **Audio**: These can be background clips or full songs (representative file types: .mp3, .wav, .m4a).

- **Movies**: Video segments of any length, displayed in either standard or high definition (representative file types: .mov, .mp4).

- **Interactions**: These can be games, tutorials, simulations, and so forth (representative file types: .swf, .xap).

I mentioned that the universality of web servers lies in their use of standard web protocols. What are these? Well, the first thing to know is that a network protocol is essentially a set of rules or conventions that allows two computer systems to recognize each other and to meaningfully exchange data.

Specifically, web servers are called HTTP servers because they use the Hypertext Transfer Protocol (HTTP) as their base network protocol. Many other protocols are involved in delivering web content, of course, including Internet Protocol (IP), Transmission Control Protocol (TCP), and Address Resolution Protocol (ARP).

NOTE: LEARN MORE ABOUT PROTOCOLS

The subject of networking protocols is far too broad and deep to cover thoroughly in this book. For more information, please check out *How the Internet Works* by my Que Publishing colleague Preston Gralla (http://is.gd/WZyOnb). In contrast to the rapid rate of change with most technologies, the Internet, for the most part, works the same way today as it did when this book was published in 2006.

Okay, then. Now that we understand a bit about what web servers are, why are they so important to us as Raspberry Pi hackers?

To answer that question, think to yourself how often you turn to a web browser to get any particular electronic task done. You can use a fully-fledged browser such as Internet Explorer or Google Chrome, or you can use a line-of-business (LOB) application that makes use of web standards to retrieve content from a web server.

For instance, you can make your Raspberry Pi a web server to accomplish any of the following goals:

- WordPress blog
- Joomla content management server
- Webcam control center
- *Minecraft* server

The list of potential projects that take advantage of an HTTP server goes on and on. The bottom line is that HTTP is a lightweight and convenient way to present online content, so why not make use of this wonderful, extensible platform on your Raspberry Pi devices?

What Is the LAMP Stack?

In web development terminology, a protocol stack is a suite of related networking protocols and technologies that fit together like finger in glove to accomplish particular kinds of work. There was a time, not too many years ago, when websites were simply static collections of manually created HTML web pages.

On the off chance this escaped your notice, I'm here to tell you that static websites have largely gone the way of the dodo. Nowadays, any web developer or designer worth his or her salt needs dynamically generated web pages that pull data from a database such as MySQL, Oracle, or Microsoft SQL Server. These dynamic websites are called data-driven web applications.

I whipped up a schematic diagram of web communications in Figure 15.1 that I hope makes this situation clearer.

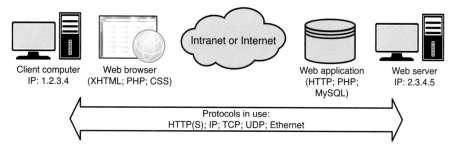

FIGURE 15.1 The basic elements of web communications

In Linux, the reference standard for an open source web development protocol stack is LAMP. LAMP in this context is an acronym for

- **Linux**: This is the base operating system for the web server.
- **Apache**: This is the world-standard open source HTTP server software.
- **MySQL**: This is the world-standard semi-open source Structured Query Language (SQL) relational database management system (RDBMS).
- **PHP**: This is the world-standard open source web development programming language. Incidentally, PHP is a strange acronym that stands for Hypertext Pre Processor.

If all this sounds like Greek to you, don't get too stressed out, now—I know I am throwing a lot of technologies and acronyms at you. Let's just take things one step at a time, and more will become clear, including how all this relates to your use of the Pi.

Installing Your Web Server

In terms of HTTP Server software, these are the major players in the world as of spring 2013 according to Netcraft (http://is.gd/uIbG30):

- Apache, by the Apache Software Foundation: http://is.gd/x3xZvH
- Internet Information Services (IIS), by Microsoft: http://is.gd/ILyw06
- nginx (pronounced *engine ex*), by Igor Sysoev: http://is.gd/ydUSGd

Of the preceding software, only IIS is proprietary. In addition, IIS is the only web server software that is platform-dependent. By contrast, Apache and nginx have software variants that run on Windows, OS X, and Linux.

NOTE: ETYMOLOGY OF APACHE

Officially, the name Apache in Apache HTTP Server was chosen out of respect for the Native American tribe of the same name. Unofficially, some people submit that Apache stands for "A Patchy," as in "Apache is continually patched and updated; therefore, it is a patchy server."

In this book, we standardize on Apache 2 as our web server of choice for the Raspberry Pi. To that point, if you're concerned that Apache might be a bit too heavy for your Raspberry Pi, lighter-weight Apache distributions are available. For instance, check out the Cherokee project (http://is.gd/IpGFx3). Frankly, I was going to use Cherokee for this chapter, but I don't feel the software is quite stable enough to recommend to you yet.

Lighttpd (pronounced *lightly*) is another example of a quality, lightweight web server. Visit the project home page at http://is.gd/pVbk3P.

TASK: INSTALLING THE LAMP STACK ON RASPBERRY PI

By setting up your Raspberry Pi as a LAMP-based web platform, you will gain an intimate understanding of how web servers work under the proverbial hood. If nothing else, you have some insider information with which you can impress your friends at the bar!

1. Open up a Terminal session and install Apache, PHP, and the library that links the two technologies together under Linux:

```
sudo apt-get install apache2 php5 libapache2-mod-php5
```

2. After the installation completes, restart the Apache service (also called daemon, pronounced *dee-mun*):

```
sudo service apache2 restart
```

3. Make sure that the default Home page was created. You can use cat to display text file contents directly onscreen:

```
cat /var/www/index.html
```

If you see some HTML tags show up (for instance, <h1>It works!</h1>, you are good to go so far.

4. Type **startx** to get into LXDE (alternatively, run a remote VNC section as you learned in Chapter 7, "Networking Raspberry Pi").

5. Open the Midori browser and navigate to the following URL:

```
http://localhost
```

If you see the output displayed in Figure 15.2, congratulations—you successfully installed Apache 2!

FIGURE 15.2 Ocular proof that the Apache 2 web server is running properly

TASK: VERIFYING PHP CONFIGURATION

PHP includes a built-in function called phpinfo() that is useful in determining whether we set up PHP correctly on our server. Let's see how it works.

1. Navigate to the default content directory in Apache 2:

```
cd /var/www
```

2. Create a new text file with an appropriate name:

```
sudo nano phpinformation.php
```

3. In nano, add the following line:

```
<?php phpinfo();?>
```

4. Press Ctrl+X, Y, and Enter to save the file and exit nano.

5. Log into LXDE if you aren't already there.

6. Open Midori and navigate to the following URL:

```
http://localhost/phpinformation.php
```

If you see the phpinfo() function output shown in Figure 15.3, you can rest assured that you successfully installed PHP.

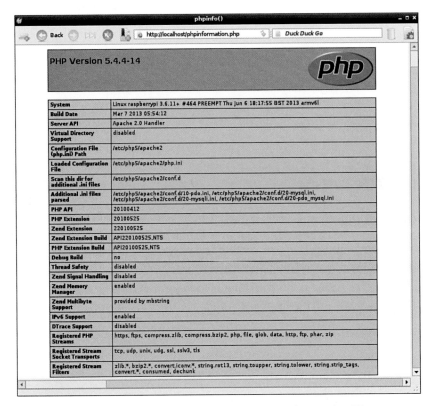

FIGURE 15.3   Verifying PHP is running correctly on the Pi

7. Now let's turn our attention to installing MySQL, the lone remaining member of the LAMP stack:

```
sudo apt-get install mysql-server mysql-client php5-mysql
```

MySQL (pronounced *my ess-cue-el*) is a client-server application (hence the necessity to install the server and client components) that is known for being fast, lightweight, and reasonably secure and extensible.

During the MySQL installation, you are prompted to set a strong password for the MySQL root user. Please do this! In my opinion, a strong password consists of the following attributes:

■ Length of at least eight characters

■ Mixture of uppercase and lowercase letters

■ Mixture of letters, numbers, and non-alphanumeric character

■ Doesn't appear in a dictionary in any language

TASK: VERIFYING MYSQL INSTALLATION

Unfortunately, MySQL does not include a quickie diagnostic function like PHP does. Here you simply check to see whether the MySQL service is present and available on your Raspberry Pi computer.

1. From a Terminal session, run the following statement:

```
sudo /etc/init.d/mysql status
```

In Linux, init.d is a directory, not a file, that contains startup and shutdown scripts for installed services on your system.

2. If you see output that resembles what's shown in Figure 15.4, you know you have a fully functioning instance of MySQL.

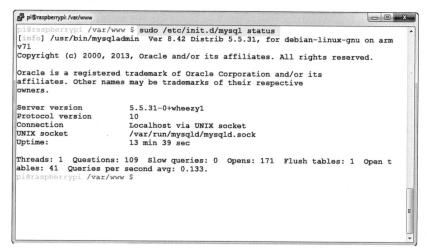

FIGURE 15.4 MySQL is installed and running on the Pi.

If you don't see the output shown in Figure 15.4, don't panic. In a worst possible case scenario, you can re-flash your SD card with Raspbian and start over from scratch. Failing that extreme measure, you can turn to the good folks at the Raspberry Pi Forums for assistance. Here is a link to a discussion thread that covers just this topic: http://is.gd/C6iONe.

Tweaking Up Your Web Server Settings

Awesome—you have your LAMP stack installed, albeit with from the factory defaults. Next, you need to get in there and make sure that the software is configured to your liking.

Apache is considered by many to be the world's best web server for many reasons, but not the least of which is that the server ships with strong and secure default values. Nonetheless, some of the most common Apache tweaks that some admins make to their default installations include

- Changing the location where web content files are stored
- Changing the default TCP port
- Modifying security and performance settings
- Adding module packages to extend the capabilities of the server

As it happens, Apache 2 stores its configuration files in the directory /etc/apache2. Specifically, the primary Apache2 configuration file is named apache2.conf, and stores general configuration parameters. Another key Apache 2 configuration file is ports.conf, located in the same directory. The ports.conf file stores TCP/IP connection settings.

You can use any text editor you want (for instance, nano from the Terminal or Leafpad from LXDE) to edit the files.

MySQL stores its configuration settings in /etc/mysql/my.cnf.

PHP stores its settings in /etc/php5/apache2/php.ini.

Transferring Content to Your Web Server

When it comes to actually authoring your website, workflows vary among developers. For static websites, all you truly need at minimum is a plain text editor to create your HTML files. For data-driven applications, such as PHP apps that read from and write to a MySQL database, a more comprehensive web authoring tool (or tools) might be more relevant.

Certainly, the subject of web development and design in itself is far outside the scope of this book. For now, let's focus on the easiest way to transfer web content from your development workstation (which I presume for now is not your Raspberry Pi) to the Pi itself.

Recall that Secure Shell (SSH) is enabled by default in Raspbian. Therefore, you can make use of the remote file-copy functionality, called Secure File Transfer Protocol (SFTP), that is part of the SSH standard to transfer your content to the Pi. This method is admirable because of its security; all session data over SSH is encrypted and consequently safe from malicious individuals.

Back in Chapter 7, I recommended FileZilla (http://is.gd/etsJLy) as a stable, reliable (and free) SFTP client. Recall that FileZilla is available on Windows, OS X, and Linux. Let's learn how to use FileZilla to move web files from a remote host to your Raspberry Pi web server.

TASK: USING SFTP TO TRANSFER CONTENT TO YOUR PI

1. Open FileZilla and click File, Site Manager to open the Site Manager tool.

2. In Site Manager, click New Site and give the connection an appropriate name.

3. On the General tab (shown in Figure 15.5), fill in the relevant connection details, like so:
- **Host**: This is the IP address of your Pi.
- **Protocol**: Select SFTP–SSH File Transfer Protocol.
- **Logon Type**: Set to Normal.
- **User**: Specify the user (pi, or another Raspberry Pi user if you have one).
- **Password**: Specify the current password for the chosen account.

4. Click Connect to save your connection and attempt to reach your Raspberry Pi. In the future you can use your stored Site Manager entry to make it convenient to reconnect to your Pi.

5. Navigate to the /var/www directory on your Pi as shown in Figure 15.5.

6. Drag and drop any content into the target Pi directory in FileZilla.

7. Click the Disconnect button on the toolbar to end your session.

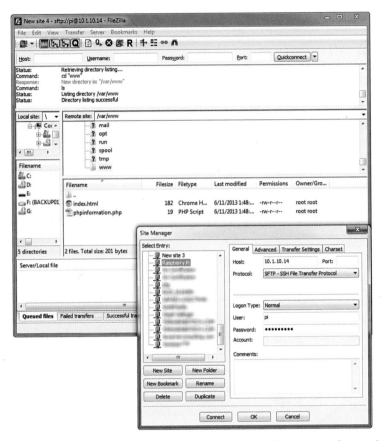

FIGURE 15.5   You can use FileZilla and SFTP to transfer web content to your Raspberry Pi.

Some long-time web users insist on using traditional FTP for transferring web content. FTP is cool because it is ubiquitous, but it has a nasty downfall: All data transmitted between the FTP server and the FTP client is clear text. That includes passwords and any other sensitive data!

Therefore, I strongly suggest you stick to using SFTP, as it employs the same command set as unencrypted FTP, has no noticeable performance penalty, and is already enabled on the Pi.

If you insist on investigating an FTP solution for your Pi, I recommend you go with vsftpd (http://is.gd/9RCFch).

TASK: INSTALL AND TEST FTP ON YOUR RASPBERRY PI

1. Install the software.

```
sudo apt-get install vsftpd
```

2. When installation completes, open the vsftpd configuration file:

```
sudo nano /etc/vsftpd.conf
```

3. Uncomment the following lines by removing the hash (#) character:

```
Anonymous_enable=NO
Local_enable=YES
Write_enable=YES
Ascii_upload_enable=YES
Ascii_download_enable=YES
```

4. Press Ctrl+O to save and then press Ctrl+X to exit nano.

5. "Bounce" or restart the vsftpd service:

```
sudo /etc/init.d/vsftpd restart
```

You can test that the FTP server works by firing up FileZilla and connecting to your Pi, specifying the FTP - File Transfer Protocol option in the Site Manager.

Alrighty then! Now that you have verified your Raspberry Pi web server is fully functional and you understand how to manually populate content, let's use a couple representative example web apps as a case study in discerning what a Raspberry Pi web server is capable of.

Setting Up phpMyAdmin

If you've had a chance to play with MySQL at all to this point, you've discovered that MySQL does not include any graphical management tools by default. As it happens, Oracle does provide a GUI toolkit called MySQL Workbench (http://is.gd/PIQrpJ). However, these Java-based tools are considered by most to be too resource-intensive for the Raspberry Pi.

Thus, kind and gentle reader, I introduce you to phpMyAdmin. phpMyAdmin (http://is.gd/T17bRC) is an open source PHP web application that provides you with a graphical front-end interface to MySQL (see Figure 15.6). Because phpMyAdmin is a web browser–based tool, you can run it on the Raspberry Pi with little to no performance impact.

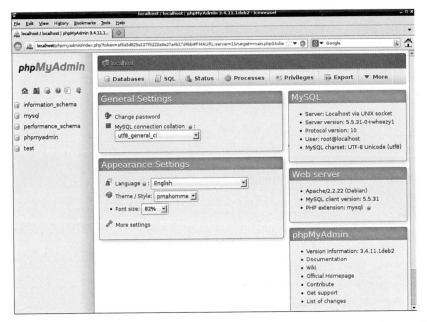

FIGURE 15.6 phpMyAdmin, a web-based MySQL administration tool

TASK: INSTALLING PHPMYADMIN

1. As usual, you need to download the software from the Raspberry Pi repositories:

```
apt-get install phpmyadmin
```

2. In the Configuring phpmyadmin screen that appears, press the Spacebar to place a selection asterisk next to apache2, which is your installed web server. Then press Tab and Enter to continue.

3. When the next Configuring phpmyadmin screen displays, select Yes to install the phpmyadmin database.

4. Type in the MySQL root user password and press Enter to continue. You'll next be asked to create and confirm a phpMyAdmin administrator password.

5. Now you need to open the Apache 2 configuration file and link Apache to phpMyAdmin.

```
sudo nano /etc/apache2/apache2.conf
```

6. In nano, press Ctrl+V repeatedly to scroll to the bottom of the file. When you are there, add the following line, save changes, and exit nano.

```
Include /etc/phpmyadmin/apache.conf
```

7. Restart Apache.

```
sudo /etc/init.d/apache2 restart
```

8. To test phpMyAdmin, start LXDE, open Midori, and navigate to the following URL:

```
http://localhost/phpmyadmin
```

Log in with the username root and whatever password you specified for the MySQL administrator. If you see the interface that is shown in Figure 15.6, you're home free!

NOTE: FOR FURTHER LEARNING

Packt Publishing has released a series of books on how to use phpMyAdmin; check them out at http://is.gd/OyFwOJ.

You know, as much as I like the idea behind Midori as a minimalist web browser, I haven't had much luck running anything but the most bare-bones of web apps from this tool. For instance, phpMyAdmin displays all these distressing artifacts on screen.

As an alternative, you might want to consider installing Iceweasel (http://is.gd/cfmCHP), the Debian port of the Mozilla Firefox web browser. Run sudo apt-get install iceweasel, confirm the installation, and check the Internet folder in LXPanel menu—you might love it!

Setting Up Joomla

Joomla (http://is.gd/Xl2hSu) is a leading content management system (CMS) platform. Many businesses build their corporate websites on Joomla because Joomla is open source, free, and eminently flexible. Under the hood, Joomla is a PHP/MySQL-based web application, so it functions perfectly well in a LAMP stack environment. I show you the Joomla default Home page in Figure 15.7.

You can also use Joomla as a blog or as an online photo/video gallery...the list is almost limitless.

NOTE: MORE ON JOOMLA!

For step-by-step instructions on how to use Joomla to build dynamic websites, please read *The Official Joomla! Book* by Jennifer Marriott and Elin Waring (http://is.gd/BA3jtO).

FIGURE 15.7 Joomla is an awesome content management platform.

Despite its richness and robustness, Joomla runs reasonably well on the Raspberry Pi. Let's learn how to install the platform.

TASK: INSTALLING JOOMLA ON YOUR RASPBERRY PI

1. From LXDE, fire up your favorite web browser, visit http://is.gd/spdPUN, and download the latest version of Joomla. The installer will come down as a ZIP archive to your home directory by default.

2. Open a Terminal session and unpack the Joomla contents to your default Apache content directory.

```
cd
sudo unzip joomla.zip -d /var/www
```

In the previous code, replace *joomla.zip* with the actual name of the Joomla ZIP you downloaded from the Joomla website.

3. You need to tweak the PHP configuration file a bit, so open it up in nano:

```
sudo nano /etc/php5/apache2/php.ini
```

4. In nano, press Ctrl+W to search for the string output_buffering.

5. Set the Development Value and Production Value parameters to 0.

6. Press Ctrl+O, ENTER, and then Ctrl+X to save the file and exit the nano editor.

7. Reboot the Pi.

```
sudo reboot
```

8. When you're back from the reboot and in LXDE again, open another Terminal session, create the Joomla configuration file, and make sure that the new file is writable.

```
cd /var/www
sudo touch configuration.php
sudo chmod 777 configuration.php
```

The touch command is used to create a new empty file. The chmod (pronounced *see aich mod*) is used to edit permissions on files. You can learn more about the Linux file system permissions, including the octal numeric and symbolic methods, by visiting good ol' Wikipedia at http://is.gd/5hFhgO.

9. It's time to complete the installation via a web browser. Open Midori, Iceweasel, or your preferred web browser and open your Apache installation's default content page:

```
http://localhost
```

10. If you see the default Apache page instead of a Joomla page, delete the old index.html page:

```
sudo rm index.html
```

11. You are prompted to walk through a three-step initial configuration wizard, the first screen of which is shown to you in Figure 15.8.

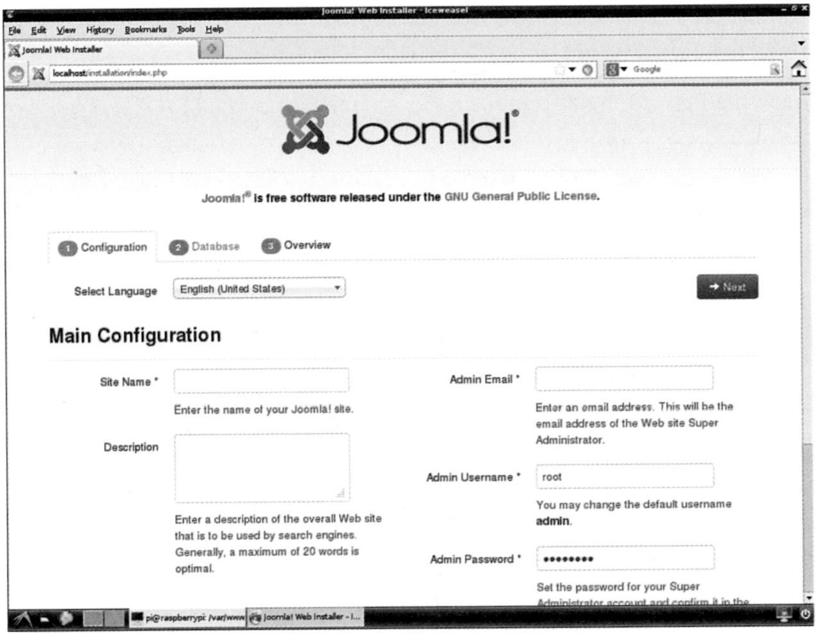

FIGURE 15.8   Joomla has a simple initial configuration wizard.

Here's a brief discussion of the information you need to supply to Joomla:

- **Main Configuration**: Site name and description; Joomla administrator login and contact details
- **Database**: Connection details to MySQL
- **Finalization**: Install sample data, email data, confirm installation defaults

12. For security purposes, you are prompted to delete the installation folder before you can begin using Joomla on your Pi. If you receive an error when you try to do this from a browser, you can perform the action through Terminal:

```
cd /var/www
sudo rm -rf installation/
```

Putting Your Web Server on the Public Internet

The final subject I cover in this chapter is how to put your Raspberry Pi web server on the global Internet. Inside the vast majority of private homes and businesses, computers use private, nonroutable IP addresses dispensed by a DHCP server. These internal IP addresses are fine for communications within the home or organization, but they don't allow people on the Internet to connect directly to those hosts.

Why would you want to expose your Raspberry Pi to the wild and wooly jungle called the World Wide Web? Here are some valid reasons:

- You can consume public web services such as Dropbox and Spotify.
- You can communicate with other Internet users.
- You can test out location services and other Internet-dependent applications.

On the other hand, you need to be mindful of some clear and present dangers associated with placing any computer within reach of systems located around the globe:

- A malicious user or application can infiltrate your Raspberry Pi.
- You may unintentionally expose private data.
- You may unknowingly consume network bandwidth (relevant for users with metered Internet connections).

Thus, my suggested workflow for putting your Pi on the Internet is to

- Configure your Pi with a static IP address. (I showed you how to do this in Chapter 7).
- Use a dynamic DNS service.

Dynamic DNS services are necessary because DHCP is a lease-based protocol. In other words, your host computers periodically receive new and different IP addresses from their servers, which makes reliable connections to computers unreliable.

Configuring your Raspberry Pi with a static private IP address is fine; this means you'll always be able to connect to the device from within your LAN by using that IP address. Dynamic DNS services come into play because they allow you to map an internal network device with a public Domain Name System (DNS) name.

In my experience, the two major players in the dynamic DNS space are

- No-IP (http://is.gd/hhmpFu)
- DynDNS (http://is.gd/nKysbj)

Both of these services offer entry-level features for free and more advanced capabilities for a subscription fee.

First, you need to visit the No-IP website and create a free account. After you've done that, you can manage your host/domain name mappings. By default, your DNS host names will use the suffix no-ip.biz. If you own a domain of your own, you can become a paying subscriber to access those additional features.

Second, you need to determine whether your Pi connects directly to your ISP and has a public IP address or if the device resides behind your router and receives a private IP.

The former case is the easiest; you see your ISP-given public IP and associated default hostname in your No-IP control panel as shown in Figure 15.9.

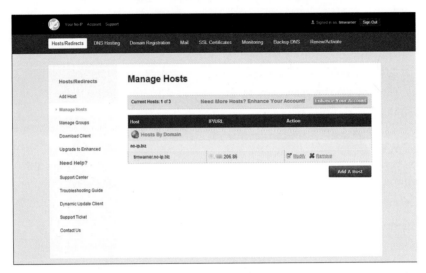

FIGURE 15.9 You can adjust your host-DNS name mappings in the No-IP Control Panel.

If your Pi is one of several hosts behind your single Internet connection, don't sweat it because the No-IP client that you'll install on your Pi is intelligent enough to sort it all out.

TASK: MAKING YOUR RASPBERRY PI PUBLICLY ACCESSIBLE BY USING NO-IP

1. Create a subdirectory inside your home directory to place the No-IP client software:

```
cd
mkdir noip
cd noip
```

2. Download the No-IP dynamic update client (DUC) software:

```
wget http://www.no-ip.com/client/linux/noip-duc-linux.tar.gz
```

3. Unpack the compressed tarball archive and navigate into the new folder:

```
ar vzxf no-ip-duc-linux.tar.gz
ls no*
cd noip-2.1.9-1
```

NOTE: VERSION CONTROL

In step 3, make sure to use the ls command to verify the name of the extracted directory. Your No-IP client version might be more recent than the one I used at the time of this writing.

4. The files you downloaded are the uncompiled source, which often comes as a surprise to Windows or OS X users that don't typically have to deal with compiling downloaded software. Enter the following commands to manually compile the software:

```
sudo make
sudo make install
```

You are prompted to enter your no-ip.com membership credentials during the client installation process on the Pi. You are also asked to specify a default refresh interval, which synchronizes your computer's IP address with the No-IP hostname.

5. Mission accomplished! You can now run the No-IP client:

```
sudo /usr/local/bin/noip2
```

If your Raspberry Pi exists as a DHCP client on your internet network, you have one more step to do. You need to log into your router and set up port forwarding to allow traffic on TCP port 80 (HTTP) to transit to your Raspberry Pi's internal IP address. The setup on my Comcast IP Business Gateway is shown in Figure 15.10.

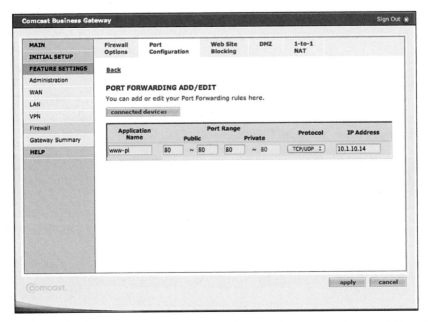

FIGURE 15.10 I need to set up port forwarding to my internal Raspberry Pi.

To test that the public IP works, fire up a web browser on another computer, preferably on a remote network, and see if you can load the Joomla site you just created. For instance, my No-IP DNS name is timwarner.no-ip.biz, so in my browser I type

```
http://timwarner.no-ip.biz
```

You will be unpleasantly surprised, I'm sure, to learn that the No-IP dynamic update client does not run automatically at startup by default. Never fear, however. You can find an excellent, step-by-step tutorial for doing this at the Stuff About Code website (http://is.gd/WolQr).

## CHAPTER 16

Raspberry Pi Portable Webcam

In this chapter you learn how to take still pictures and record video with your Raspberry Pi. Perhaps you want to investigate time lapse photography, install a baby monitor, set up a security camera, deploy a bird feeder cam, or simply snap interesting images.

Prior to May 2013, when the Raspberry Pi Foundation introduced the Raspberry Pi Camera Board, enthusiasts did their best to use their USB webcams with their Raspberry Pi units. To be sure, this chapter teaches you how to use the Pi with third-party cameras. However, I must tell you that the Raspberry Pi Camera Board is pretty slick!

I conclude this lesson with step-by-step instructions and best practice advice for putting your Raspberry Pi on battery and therefore freeing you up to take your new webcam wherever you need to take it. Let's get started!

About the Raspberry Pi Camera Board

As you know, the Model A and Model B boards include a Camera Serial Interface 2 (CSI-2) camera connector, shown in Figure 16.1. The interface is labeled S5 and is located between the USB and HDMI ports on the Pi PCB.

FIGURE 16.1   The MIPI CS-2 camera interface on a Model A board

Initially the Foundation said nothing as to whether it would create a camera to connect to this interface, leaving Raspberry Pi enthusiasts to speculate as to how they could access the CSI-2 interface through hardware hacking. Other Pi users simply plugged in their USB-connected cameras; we cover that idea later in this chapter.

This situation cleared up in May 2013 when the Raspberry Pi Foundation announced a $25 accessory called the Raspberry Pi Camera Board, which does in fact connect to the Pi through the CSI-2 interface.

The Raspberry Pi camera board is available through the typical channels:

- **RS Components**: http://is.gd/6ol2Gq
- **Premier Farnell/Element 14**: http://is.gd/xJSkbQ

Physically, the CSI connector (the one shown in Figure 16.1) implements a 15-pin flex ribbon cable. One end connects to the CSI-2 interface on the Raspberry Pi PCB. The other end of that cable is soldered directly to the Raspberry Pi Camera Board, as shown in Figure 16.2.

FIGURE 16.2   Raspberry Pi Camera Board PCB

The specs for this board (also called *Camera Module*) are listed in Table 16.1.

TABLE 16.1   Raspberry Pi Camera Board Specifications

Dimensions	25mm x 20mm x 9mm
Weight	3g
Sensor Make and Model	Omnivision OV5647
Sensor Resolution	5 megapixels (MP)
Focus Type	Fixed
Fixed Picture Resolution	2592x1944 pixels (px)
Video Resolutions	1080p @ 30 frames per second (fps); 720p @ 60fps; 640x480 @ 60 or 90fps
Flash	None
Power Consumption	100mA at 1.5V

I think you'll find that the Raspberry Pi camera board specs compare favorably with those of, say, the iPhone 4 from summer 2010. Not too shabby! Actually, I found that the Camera Board shoots better video than many entry-level webcams I've used in the past.

The Camera Board ships in an anti-static bag enclosed by a minimalist paper box. When you handle the Camera Board, be careful not to kink the ribbon cable—it is on the delicate side. You also want to avoid touching the camera lens to avoid fingerprint smudges.

With no further ado, let's get your new Raspberry Pi Camera Board set up and start snapping some pictures and recording some video!

Installing and Configuring the Raspberry Pi Camera Board

Getting the Raspberry Pi Camera Board up and running consists of two phases:

- Preparing the Pi and installing the camera driver
- Physically installing the camera

As usual, in this chapter I assume you're using the official Raspbian Linux distribution.

TASK: PREPARING THE RASPBERRY PI FOR THE CAMERA BOARD

It is never a good idea to connect hardware to a computer's motherboard while that computer is powered up. That said, before you unplug the Pi and physically attach the Camera Board, you need to enable the use of the camera in Raspbian.

As you'll learn the raspi-config utility provides a simple interface for managing the camera functionality. Let's do that now.

1. Do not plug in the Camera Board yet. Power on the Pi, access a Terminal prompt, and update your software (including Raspi-Config):

```
sudo apt-get update && sudo apt-get upgrade
```

2. Let's go into Raspi-Config:

```
sudo raspi-config
```

3. In Raspi-Config, arrow down to the Camera option and press Enter.

4. In the Enable support for Raspberry Pi camera? dialog box, shown in Figure 16.3, select Enable and press Enter.

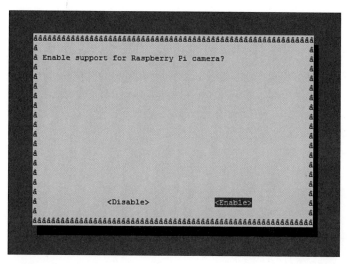

FIGURE 16.3   Enable the Camera Board by using Raspi-Config.

5. Exit the Raspi-Config utility and reboot your Pi. When you are back from the reboot, shut down the system to prepare for the Camera Board installation. Remember that the shutdown command uses two primary switches: -h for shutdown (halt), and -r for reboot.

```
sudo shutdown -h now
```

Now that you have installed the necessary device drivers and readied the Raspberry Pi to use the Camera Board, let's connect the add-on to the Pi's PCB.

TASK: INSTALLING THE RASPBERRY PI CAMERA BOARD

Okay. Now that you've notified your Pi's software that you want to use the Camera Board, and you've removed power from the Pi (is that similar to "power to the people"? Never mind...), you can proceed with the physical installation.

One preliminary word of caution: The CSI-2 interface on the Raspberry Pi is delicate. Don't use too much force or you may break the retaining clips and render the entire interface useless. Let's get to work!

1. Make sure that your Raspberry Pi is powered off. Unplug all cables from the PCB.

2. Using your fingers, grasp the edges of the CSI-2 connector and gently lift up the retaining clip. Note that the clip remains attached to the interface; it lifts approximately 1–2mm.

3. Insert the Camera Board ribbon cable into the CSI-2 interface with the copper traces facing away from the USB ports. You can see the correct orientation in Figure 16.4.

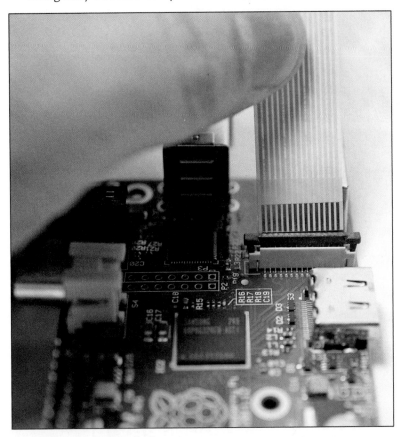

FIGURE 16.4 The Camera Board installation procedure is a bit tedious, and the components are certainly delicate.

4. When the ribbon cable is seated in the interface, grasp the retaining clip with your fingers and gently press down to lock the cable and the interface together securely.

Alrighty then! As you can see in Figure 16.5, the Raspberry Pi and the Camera Board are now (hopefully) a functional unit.

FIGURE 16.5   The Raspberry Pi and the Camera Board add-on make a nice pair, don't they?

Using the Camera Board

The Raspberry Pi Foundation provides two command-line utilities for using the Camera Board:

- **raspistill**: Used to take still images in both JPEG and RAW formats
- **raspivid**: Used to record video by using the H.264 codec

In the next section I cover how to use raspistill to take still pictures. In the section that follows I turn your attention to shooting full-motion video (FMV) using raspivid.

Capturing Still Pictures

Let's begin by obtaining some command-line syntax help:

```
raspistill | less
```

Figure 16.6 shows you the screen output for raspistill; this should serve as a nice reference for you.

```
                                        pi@raspberrypi ~                          _ □ x
File  Edit  Tabs  Help
pi@raspberrypi ~ $ raspistill | less

raspistill Camera App v1.1

Runs camera for specific time, and take JPG capture at end if requested

usage: raspistill [options]

Image parameter commands

-?, --help       : This help information
-w, --width      : Set image width <size>
-h, --height     : Set image height <size>
-q, --quality    : Set jpeg quality <0 to 100>
-r, --raw        : Add raw bayer data to jpeg metadata
-o, --output     : Output filename <filename> (to write to stdout, use '-o -'). If not specified, no file is sav
ed
-v, --verbose    : Output verbose information during run
-t, --timeout    : Time (in ms) before takes picture and shuts down (if not specified, set to 5s)
-th, --thumb     : Set thumbnail parameters (x:y:quality)
-d, --demo       : Run a demo mode (cycle through range of camera options, no capture)
-e, --encoding   : Encoding to use for output file (jpg, bmp, gif, png)
-x, --exif       : EXIF tag to apply to captures (format as 'key=value')
-tl, --timelapse          : Timelapse mode. Takes a picture every <t>ms

Preview parameter commands

-p, --preview    : Preview window settings <'x,y,w,h'>
-f, --fullscreen          : Fullscreen preview mode
-op, --opacity   : Preview window opacity (0-255)
-n, --nopreview  : Do not display a preview window

Image parameter commands

-sh, --sharpness          : Set image sharpness (-100 to 100)
-co, --contrast  : Set image contrast (-100 to 100)
-br, --brightness         : Set image brightness (0 to 100)
-sa, --saturation         : Set image saturation (-100 to 100)
-ISO, --ISO      : Set capture ISO
-vs, --vstab     : Turn on video stabilisation
-ev, --ev        : Set EV compensation
-ex, --exposure  : Set exposure mode (see Notes)
-awb, --awb      : Set AWB mode (see Notes)
-ifx, --imxfx    : Set image effect (see Notes)
-cfx, --colfx    : Set colour effect (U:V)
-mm, --metering  : Set metering mode (see Notes)
-rot, --rotation          : Set image rotation (0-359)
-hf, --hflip     : Set horizontal flip
-vf, --vflip     : Set vertical flip


Notes

Exposure mode options :
off,auto,night,nightpreview,backlight,spotlight,sports,snow,beach,verylong,fixedfps,antishake,fireworks

AWB mode options :
off,auto,sun,cloud,shade,tungsten,fluorescent,incandescent,flash,horizon

Image Effect mode options :
none,negative,solarise,sketch,denoise,emboss,oilpaint,hatch,gpen,pastel,watercolour,film,blur,saturation,colour
swap,washedout,posterise,colourpoint,colourbalance,cartoon
```

FIGURE 16.6 raspistill command syntax

You can also download the full documentation for the Camera Board commands from http://is.gd/18PvNf.

To tell your Pi to snap a picture, enter the following command:

```
raspistill -o myimage.jpg
```

When you run raspistill, you see an LED light up on the Camera Board for approximately four seconds; the image is exposed just before the LED goes out.

Any photographs you capture are stored in your present working directory. Thus, if you execute the raspistill command in the context of your home directory, that's where your files reside by default.

The -o switch enables you to name the images using whatever file name you input after the switch, myimage.jpg in this example.

From LXDE, you can double-click the image files to open them in your default web browser. Alternatively, you can right-click them and select ImageMagick (display) to open them in the ImageMagick open source image viewer.

Despite the lack of a flash or manual focus override, the picture quality is actually pretty good. Take a look at Figure 16.7 to see yours truly posing for your viewing (and laughing) pleasure:

FIGURE 16.7 Despite the poor subject, you can see that the Raspberry Pi Camera Board takes a pretty good picture.

Let's buzz through some more sample syntax to give you a better idea as to what raspistill can do for you. You can, for example, tell your Pi's camera to take a picture at a set delay or image quality.

Take an image with a quality of 50% and a "shutter" delay of 10 seconds:

```
raspistill -o image3.jpg -q 50 -t 10000
```

The -q parameter goes from 0 (lowest quality) to 100 (highest quality). Quality, in this case, refers to the degree of JPEG compression that is applied to captured images. JPEG is a lossy compression algorithm, so even images taken at quality 100 will have some pixel loss due to the file format.

The timer value adds an exposure delay and works in thousands of seconds (milliseconds). Thus, a value of 5000 represents a 5 second delay.

Take an image with custom dimensions, verbose command output, a quality of 80%, and a one-second delay:

```
raspistill -v -w 1024  -h 768 -q 80 -o image4.jpg -t 1000
```

The verbose (-v) parameter is useful for educational and troubleshooting purposes. In fact, let me show you the output of the previous raspistill command example:

```
pi@raspberrypi ~ $  raspistill -v -w 1024 -h 768 -q 80 -o image4.jpg -t 1000 >
output.txt


raspistill Camera App v1.2


Width 1024, Height 768, quality 80, filename image4.jpg
Time delay 1000, Raw no
Thumbnail enabled Yes, width 64, height 48, quality 35
Full resolution preview No


Preview Yes, Full screen Yes
Preview window 0,0,1024,768
Opacity 255
Sharpness 0, Contrast 0, Brightness 50
Saturation 0, ISO 400, Video Stabilisation No, Exposure compensation 0
Exposure Mode 'auto', AWB Mode 'auto', Image Effect 'none'
Metering Mode 'average', Colour Effect Enabled No with U = 128, V = 128
Rotation 0, hflip No, vflip No
ROI x 0.000000, y 0.000000, w 1.000000 h 1.000000
Camera component done
Encoder component done
Starting component connection stage
Connecting camera stills port to encoder input port
Opening output file image4.jpg
Enabling encoder output port
Starting capture 1
Finished capture 1
Closing down
Close down completed, all components disconnected, disabled and destroyed
```

Although the previous verbose output looks like so much gobbledygook at first glance, given experience and practice you'll come to appreciate the degree of detail that raspistill gives you. The verbose output leaves nothing to the imagination, so you can figure out the source of any unexpected behavior you see in taking still pictures with your Raspberry Pi.

NOTE: ABOUT FLASH

Because neither the Raspberry Pi Camera Board nor any webcam I've ever used includes a built-in flash, you need to pay attention to ambient light when you capture still pictures or video. If you visit the Raspberry Pi forums (check out these threads: http://is.gd/LaKy6m; http://is.gd/d1SoH5) you'll find some enthusiastic debate regarding the relative difficulty of accessing the Pi's GPIO pins to attach and sync a flash bulb with the camera sensor.

The consensus among Pi developers at the time of this writing is that the Omnivision Camera Board firmware does include support for a flash, but digging into that firmware source code isn't something that the Raspberry Pi Foundation has addressed yet.

Recording Video with the Camera Board

As I mentioned earlier, the Raspberry Pi Foundation gives you the raspivid command-line utility to capture full-motion video on your Pi.

Let's get right into the thick of things by learning the raspivid command syntax:

```
raspivid | less
```

For those of you who want to see the raspivid syntax right now (nothing like instant gratification, right?), let me give you the results of raspivid --help.

```
pi@raspberrypi ~ $ raspivid --help
Display camera output to display, and optionally saves an H264 capture at
requested bitrate
usage: raspivid [options]
Image parameter commands
-?, --help      : This help information
-w, --width     : Set image width <size>. Default 1920
-h, --height    : Set image height <size>. Default 1080
-b, --bitrate   : Set bitrate. Use bits per second (e.g. 10MBits/s would be -b
10000000)
-o, --output    : Output filename <filename> (to write to stdout, use '-o -')
-v, --verbose   : Output verbose information during run
-t, --timeout   : Time (in ms) to capture for. If not specified, set to 5s. Zero
to disable
-d, --demo      : Run a demo mode (cycle through range of camera options, no
capture)
-fps, --framerate      : Specify the frames per second to record
-e, --penc      : Display preview image *after* encoding (shows compression
artifacts)
-g, --intra     : Specify the intra refresh period (key frame rate/GoP size)
Preview parameter commands
```

```
-p, --preview    : Preview window settings <'x,y,w,h'>
-f, --fullscreen      : Fullscreen preview mode
-op, --opacity  : Preview window opacity (0-255)
-n, --nopreview : Do not display a preview window
Image parameter commands
-sh, --sharpness        : Set image sharpness (-100 to 100)
-co, --contrast : Set image contrast (-100 to 100)
-br, --brightness       : Set image brightness (0 to 100)
-sa, --saturation       : Set image saturation (-100 to 100)
-ISO, --ISO     : Set capture ISO
-vs, --vstab    : Turn on video stablisation
-ev, --ev       : Set EV compensation
-ex, --exposure : Set exposure mode (see Notes)
-awb, --awb     : Set AWB mode (see Notes)
-ifx, --imxfx   : Set image effect (see Notes)
-cfx, --colfx   : Set colour effect (U:V)
-mm, --metering : Set metering mode (see Notes)
-rot, --rotation        : Set image rotation (0-359)
-hf, --hflip    : Set horizontal flip
-vf, --vflip    : Set vertical flip
-roi, --roi     : Set region of interest (x,y,w,d as normalised coordinates
[0.0-1.0])
Notes
Exposure mode options :
off,auto,night,nightpreview,backlight,spotlight,sports,snow,beach,verylong,fixedfp
s,antishake,fireworks
AWB mode options :
off,auto,sun,cloud,shade,tungsten,fluorescent,incandescent,flash,horizon
Image Effect mode options :
none,negative,solarise,sketch,denoise,emboss,oilpaint,hatch,gpen,pastel,
watercolour,film,blur,saturation,colourswap,washedout,posterise,colourpoint,
colourbalance,cartoon
Metering Mode options :
average,spot,backlit,matrix
```

You probably noticed that the raspivid command syntax is almost identical to that of raspistill. This behavior, of course, is by design.

Now let's record a quick five-second video:

```
raspivid -o fiveseconds.h264
```

By default, the camera's video capture is full 1080p HD at 1920x1080 pixels. Of course, you can make a smaller capture; let's make a 15-second clip:

```
raspivid -o smallvid.h264 -t 15000 -w 1024 -h 768
```

NOTE: PERHAPS A LARGER SD CARD IS IN ORDER...

Recording at 1080p equates to a disk storage footprint of 17 megabytes (MB) per second, or 115MB per minute of video. Thus if you plan on capturing a significant amount of video with your Pi, you might want to purchase a higher capacity SD card.

Of course, the Raspberry Pi Camera Board has no microphone, so your videos won't have any audio. Actually, adding an audio feed to your video captures is a good Raspberry Pi project idea!

TASK: ENCODING A RASPIVID VIDEO FILE

I have some bad news and some good news for you with regard to raspivid captures. The bad news is that the capture is a raw H.264 data stream that isn't immediately viewable on the Pi or any computer, for that matter. The good news is that you can wrap the raw H.264 stream into an MPEG Layer 4 (MP4) container by following these steps:

1. Download and install MP4Box (http://is.gd/Fbti7Z):

```
sudo apt-get install -y gpac
```

The -y parameter is useful if you want to approve the download automatically instead of having to manually specify "y."

2. Remember that Linux is a completely case-sensitive operating system. Thus, you must use the command MP4Box and not mp4box, MP4box, or any combination or permutation thereof:

```
MP4Box -fps 30 -add smallvid.h264 smallvid.mp4
```

This command feeds in the smallvid.h264 movie file you created using raspivid earlier; that is the purpose of the -add parameter. It finishes with the filename of the encoded movie file; in this case, smallvid.mp4.

3. Raspbian includes the Omxplayer (http://is.gd/UnBEuD) media player that you can use to view your newly converted video capture. To use it just execute the program with the video file name:

```
omxplayer smallvid.mp4
```

You can get full documentation on Omxplayer, including keyboard controls, by visiting the Embedded Linux Wiki at http://is.gd/PNC7Mf.

If you have trouble with Omxplayer, you can try my personal favorite media player on any platform—VLC (http://is.gd/kfwzk5):

```
sudo apt-get install -y vlc
vlc smallvid.mp4
```

From LXDE, you can also right-click your MP4 file and select VLC media player from the shortcut menu.

Using a Third-Party USB Webcam

If you already have a Raspberry Pi-compatible USB webcam, feel free to use that piece of hardware instead of purchasing the Camera Board. You can check your webcam against the list of verified peripherals here at http://is.gd/ZJA79A.

For my part, I have a nifty Logitech HD Webcam C615, shown in Figure 16.8, that works on my Pi like a champ even without a powered hub!

FIGURE 16.8   I actually prefer using a third-party webcam to the Raspberry Pi Camera Board.

For this exercise we use an awesome piece of open source software called Motion.

TASK: CONFIGURING A THIRD-PARTY USB WEBCAM

One awesome thing (among many) about the Raspberry Pi is that you are rarely, if ever, locked into any particular method of accomplishing a task. Take recording webcam video, for instance. You don't have to buy the Raspberry Pi Camera Board if you already have a third-party webcam.

Let's learn how to use a standard USB webcam with the Pi.

1. Plug in your webcam, preferably into a powered USB hub.

2. From a Terminal shell prompt, make sure your Raspberry Pi recognizes your device:

   ```
   lsusb
   ```

   As you can see in Figure 16.9, my Logitech C615 is detected.

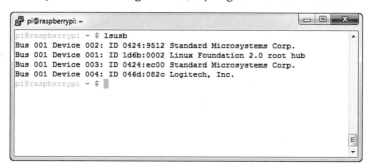

The Pi detects this Logitech webcam.

FIGURE 16.9 The lsusb command is used to enumerate USB devices in Linux. Here you can see both my Logitech webcam (device 004) as well as my Scosche portable battery (device 001).

3. To get the camera working properly with your Pi, you also need to edit two Motion configuration files a bit. To edit the first file, enter the following command:

   ```
   sudo nano /etc/default/motion
   ```

   In this configuration file, change the value

   ```
   start_motion_daemon=no
   ```

   to

   ```
   start_motion_daemon=yes
   ```

   Make sure to save your changes before you exit nano.

NOTE: SMILE—YOU ARE ON THE WORLD WIDE WEB

If you are interested in making your Raspberry Pi webcam accessible from the public Internet, look no further than Chapter 15, "Raspberry Pi Web Server," in which I give you instructions on using the lovely No-IP service.

4. Now for the second configuration file tweak:

```
sudo nano /etc/motion/motion.conf
```

Change the Daemon option from Off to On.

Change the webcam_localhost parameter from On to Off.

Again, save and exit.

5. Start the video stream:

```
sudo service motion start
```

To view your live video stream, open a web browser (I suggest Iceweasel or Netsurf [http://is.gd/6AfWdc]) and navigate to the following URL:

```
http://localhost:8081
```

If you want to access your webcam remotely on your LAN, substitute your Pi's IP address for localhost.

You can tweak detailed parameters of the stream including stuff like the following:

- Video dimensions
- Frame rate
- Video quality
- Capture storage directory

by editing /etc/motion/motion.conf. You can see my goofy mug (again) in Figure 16.10.

FIGURE 16.10   This is Motion webcam output.

When you want to stop the stream, issue the following Terminal command:

```
sudo service motion stop
```

The Pale Blue Dot blog (http://is.gd/XS6fY8) has good instructions for setting up Motion to run as a service and auto-start each time your Pi boots up. For security reasons, though, I advise you to be careful about this, especially if your Pi is publicly accessible.

You're probably wondering if the raspistill and raspivid commands that were written to function with the official Raspberry Pi Camera Board work with a third-party webcam. I've tested this out for you, and the answer is no. If you try to run raspistill or raspivid against a third-party webcam, you will receive an error message that says in part:

```
Camera is not detected. Please check carefully the camera module is installed
correctly.
```

Setting Up Your Webcam

As you just saw, Motion provides you with a video stream and a web browser in one fell swoop. But what if you wanted your Raspberry Pi cam to monitor a particular area (be it your front door, your back yard, your fish tank, whatever) and snap a picture at regular intervals?

Our current Motion setup feeds a live stream to a tiny web server on port 8080 (configurable through montion.conf, naturally).

TASK: SETTING UP A TIME-LAPSE WEBCAM

Let's say that your goal is to have the webcam snap a picture every 30 seconds and save the image snapshot files in /home/pi/webcam. To do this let's try out a different piece of webcam software: fswebcam (). Be sure to stop the Motion service by issuing **sudo service motion stop** before proceeding with the following procedure:

1. Install fswebcam:

```
sudo apt-get install fswebcam
```

2. You can get a run of the tool's command-line help:

```
fswebcam --help
```

3. Now let's snap a decent-sized image. Note that fswebcam uses the default ID /dev/video0 for the first webcam it sees; assuming you have only one webcam installed, this ID should work fine for your purposes.

```
fswebcam -r 1024 x 768 -d /dev/video0 picname.jpg
```

In the previous example, the -r switch specifies the image dimensions, and the -d switch specifies the output directory. Finally, comes the name of the output image file in JPEG format.

Of course, picname is a generic identifier for your picture file.

4. Create a configuration file for fswebcam so you can set your preferred defaults, especially your output directory:

```
cd
sudo nano .fswebcam.conf
```

5. When you are in the file, add these lines, customizing the values to your liking (I'm showing you my own setup here for illustrative purposes):

```
device /dev/video0
input 0
loop 15
skip 20
background
resolution 320x240
set brightness=60%
```

```
set contrast=13%
top-banner
title "Warner Webcam"
timestamp "%d-%m-%Y %H:%M:%S (%Z)"
jpeg 95
save /home/pi/webcam/viewcam.jpg
palette MJPEG
```

Pay particular attention to the timestamp parameter; this is where you can differentiate your captured image files as well as make them easier to browse. Note also all the options you have to customize the webcam's default behavior.

6. To start fswebcam using your new configuration file (assuming that the config file resides in the root of your home directory), type

```
fswebcam -c ~/.fswebcam.conf
```

In case you were wondering, the dot before the .fswebcam.conf makes the file hidden. This is normally the attribute that is attached to system and configuration files to keep novices from accidentally messing with them.

7. To stop the fswebcam process, issue this command:

```
pkill fswebcam
```

8. To create a repeating operating system-level job that snaps a picture every minute, turn to Bash shell scripting and the age-old Linux command cron. First you must create a shell script:

```
sudo nano camscript.sh
```

NOTE: ABOUT CRON

The cron (pronounced *krahn*) utility has been around since the earliest days of Unix and Linux. Use cron to schedule tasks to run once or on a schedule. In fact the name cron derives from the Greek word chronos, which means time. Specifically, the tasks that cron runs are typically binary commands or shell scripts.

9. Enter the following contents into the script file:

```
#!/bin/sh
filename=$(date +"%m-%d-%y|||%H%M%S")
fswebcam -r 356x292 -d /dev/video0 $filename.jpg
cp $filename.jpg /home/pi/webcam
```

I'll explain each of those four lines for you:

1: This is called a "shebang" line and points Linux to the location of the sh command interpreter.

2: This creates a variable named filename that gives a date and time stamp as its value.

3: This invokes fswebcam; in this example we aren't using a configuration file.

4: This copies the captured image (stored in the filename variable) to a subfolder. We could add additional code here to perform an upload to Dropbox, FTP transfer, and so forth.

10. Mark your new shell script as executable:

```
sudo chmod +x camscript.sh
```

11. Start the crontab editor for the pi user:

```
crontab -e
```

12. The nano editor opens. Move your insertion point to the bottom of the file and create a cron job to run the camscript.sh shell script every minute:

```
*/1 * * * * /home/pi/camscript.sh
```

The previous syntax looks strange with all the asterisks, doesn't it? In a nutshell, the Cron format uses six fields, with an asterisk representing the entire range of possible values for that field. Moving from left to right, the fields are:

■ Minute (range 0-59)

■ Hour (range 0-23)

■ Day of the Month (range 1-31)

■ Month of the Year (range 1-12)

■ Day of the Week (range 1-7, with 1 signifying Monday)

■ Year (range 1900-3000)

Thus, a Cron statement of * /1 * * * * denotes an interval of one minute with no other limitations, be they day, month, or year.

I found an outstanding article on Cron format that you'll want to have a look at if you want more information: http://is.gd/bCCmOm.

Check out Figure 16.11 to see what my setup looks like. It works like a charm!

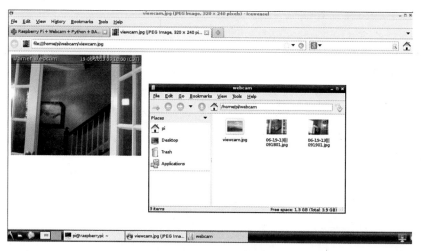

FIGURE 16.11 This is my time-lapse webcam. At left in the browser window you can see the live video feed. At right you can see my output folder superimposed; this is where the .jpg image snapshots are stored.

Adding a Rechargeable Battery Pack to Your Pi Camera

Attaching a rechargeable battery pack to your Raspberry Pi (especially when combined with a Wi-Fi dongle) makes your Pi eminently flexible. In this scenario, you could feasibly place your Pi webcam anywhere, either indoors or outdoors, and happily snap pictures, record video, or both. You can even program the Pi to send the capture files to a desktop computer by using FTP!

Raspberry Pi enthusiasts around the world have developed novel ways to provide 5-volt battery power to their Pis. Some of these ideas involve car batteries, 12V vehicle power sockets, and the like. However, we have much more basic and reliable methods available to us.

I suggest you look into a Lithium-Ion (Lion) battery pack that includes a Micro-B USB plug and regulated 5V output. This option means that (a) you can plug the Lion portable battery into your desktop computer or a powered USB hub to charge the battery; and (b) you can simply plug the Micro-B plug into your Pi's power port to give it portable juice on the go!

As usual, the Embedded Linux wiki (http://is.gd/ETvw9J) has a list of external battery packs that have been verified to work with the Raspberry Pi.

Remember that the standard power draw of the Model B board is 500mA. The amount of energy in a battery pack is ordinarily given in milliampere hours (mAh). This means that a 1,000mAh battery pack can deliver 1000mA of power for one hour, or 500mA for two hours.

Adafruit sells a Lion battery pack (http://is.gd/7ID1JP) for $59.95 that contains 3300mAh worth of power. This equates to over six hours of continuous power to a normally operating Raspberry Pi unit. Pretty cool, eh?

Some Lion portable batteries, like my own Scosche IPDBAT2 shown in Figure 16.12, have two 5V USB ports with different current supplies: 1A and 2.1A. This difference is normally intended to allow the battery pack to supply power to either the iPhone (1A current) or iPad (2.1A current). For the Pi, I suggest you go with the 2.1A port.

FIGURE 16.12   This is my Raspberry Pi battery pack: It gives me over six hours of power!

Python and Your Webcam

A good choice for blending the power of Python programming with your webcam is the SimpleCV vision library (http://is.gd/Db2osA). One aspect of SimpleCV that you might want to investigate is the motion and face detection capabilities built into the library.

TASK: SETTING UP SIMPLECV

Before you begin, make sure your USB webcam is connected, detected, and ready to rock. You should also stop any existing webcam services you might have running on your Pi and then follow these steps:

1. Install the SimpleCV libraries:

```
sudo apt-get install python-opencv python-scipy python-numpy python-pip
sudo pip install https://github.com/ingenuitas/SimpleCV/zipball/master
```

2. Verify that the SimpleCV Python 2 library loads correctly in the Python interpreter:

```
python
import SimpleCV
print dir(SimpleCV)
```

3. Press Ctrl+D to exit Python.

4. You should create a very, very simple Python 2 script to test SimpleCV's functionality. Begin by creating the script file:

```
cd
sudo nano hellocamera.py
```

5. Add the following code to the new script file:

```
#!/usr/bin/env python
from SimpleCV import Image, Display
from time import sleep

myWindow = Display()

myImage = Image("webcam.jpg")

myImage.save(myWindow)

while not myWindow.isDone():
    sleep(0.1)
```

In a nutshell, this Python 2 script performs the following actions:

- Imports relevant functions (methods) from the SimpleCV and sleep modules
- Creates a display window
- Loads a webcam snapshot to the newly created window and saves the file to the current working directory
- Prevents the script from terminating immediately after the webcam snapshot is taken

Save your work and close the file when you're finished.

6. Now let's run the Python script file from a shell prompt:

```
python hellocamera.py
```

Within moments, you should see your webcam snapshot appear on screen (see Figure 16.13). Press Ctrl+C to abort the script execution.

FIGURE 16.13   Using SimpleCV to access a third-party webcam. The goofy subject is yours truly, and the bit of artwork partially visible above his head is courtesy of the author's three-year-old daughter Zoey.

## CHAPTER 17

Raspberry Pi Security and Privacy Device

When it comes to electronics and computing, the words security and privacy could mean a lot of different things. In this case I'm talking about using your Pi to maximize your security and privacy when online. Even more specifically, I'm referring to securing your personal and financial details.

I want to lead off this chapter by presenting to you three unfortunate scenarios that can be prevented by configuring your Raspberry Pi as a security and privacy device.

Scenario #1: You live in the United States, and you enjoy your subscription to Netflix very much. However, you find that whenever you visit other countries, especially in Europe and the Far East, that you are unable to access the Netflix services you pay for due to international licensing laws.

Scenario #2: You enjoy taking your laptop computer to the neighborhood coffee house and working while sipping espresso and munching scones. The free, public Wi-Fi service gives you adequate Internet access speed. Nonetheless, you discover three months later that your bank accounts have been compromised because a hacker captured your logon credentials over the air during one of your coffee house web browsing sessions.

Scenario #3: You are a traveling salesperson who spends much of your time doing your work and browsing the Web from hotel rooms. You've become increasingly irritated at how much of the Web is blocked by hotel access restrictions.

This chapter is all about leveraging your $25 or $35 Raspberry Pi computer to provide yourself with secure and potentially anonymous Internet access. Your first order of business, of course, is to define your terms. Let's start off with the concept of the *virtual private network* and why it's important.

Encrypting Your Internet Connection with a VPN

A virtual private network, or VPN, is a secure, point-to-point network connection that is tunneled through an unsecure medium. You can certainly agree that the public Internet represents an unsecure medium; a wild jungle is more like it.

A VPN involves the construction of a temporary or permanent data communications channel that takes advantage of the Internet's speed and reliability, but offers security in that all traffic tunneled through the VPN connection is encrypted.

VPNs are the *de facto* method that businesses use to offer remote employees access to confidential, internal network resources such as shared files, intranet websites, and so forth.

If there is a downside to VPNs, it is their lack of speed. Due to the heavy overhead of data encryption and decryption, network access over a VPN pipe is noticeably, and sometimes unbearably, slower than over an unencrypted link.

That brings up the delicate balancing act between increased security on one hand and user convenience on the other. Where are you comfortable drawing the line?

From the perspective of the corporate IT manager, setting up a VPN server can be pretty tough. You ordinarily have to provision dedicated hardware and software in order to manage the myriad network protocols that constitute VPN circuits.

What many people don't know is that it is relatively easy to create your own VPN environment, even at home, by using hosted VPN services. Two major players in the hosted VPN arena are

- **LogMeIn Hamachi**: http://is.gd/PtTdkV
- **OpenVPN**: http://is.gd/uSoYCp

What is cool about both of these services is that because they operate over standard web protocols, you don't have to forward any ports on your router or worry about your Internet access provider blocking traditional VPN ports.

For instance, I've stayed at hotels that charge guests two different prices for Internet access depending on whether the guest needs VPN access or not.

However, in my experience Hamachi is much more of a turnkey solution than OpenVPN, so I focus on this product in this chapter.

Okay...so you understand that a VPN connection gives you security and privacy by protecting all data between your computer and the Internet. What does that have to do with the Raspberry Pi? Moreover, how does a VPN enable you to cloak your geographic location?

The truth of the matter is, a VPN alone cannot give you true web browsing freedom. For that purpose we need a proxy server.

Browsing Where You Want via a Proxy Server

A proxy server is a networking device that connects to Internet resources on behalf of another computer. For instance, you may find that network broadcasting agreements prevent your favorite sports game from being broadcast in your current location. Wouldn't it be cool to have your computer appear as if it were connecting from another location so you could watch your game?

NOTE: PUBLIC PROXIES AND THE NETWORK ADMINISTRATORS

Many network administrators detest public proxies because they allow users to bypass corporate web browsing filters. Businesses oftentimes set up what's called a *transparent proxy* that forces all internal network traffic through the device before it hits the Internet. To that point, many transparent proxies periodically download blacklists of known public proxies to thwart unauthorized use of the network.

Pipelining your network traffic through a proxy server offers you privacy because as far as your Internet access provider is concerned, you are making an ordinary, run-of-the-mill web browsing request to a particular server.

In actuality, that particular server is a proxy server that is capable of redirecting your web browsing to wherever you need or want to go (see Figure 17.1). It's common for citizens of certain countries to use web proxies to bypass their government's Internet access filters.

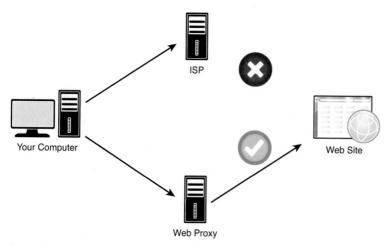

FIGURE 17.1 How a web proxy works

Let me explain what you're seeing in Figure 17.1. Your computer is unable to connect to a target website due to filtering from your Internet access provider. This "Internet access provider" could be your residential Internet Service Provider (ISP), a corporate Internet connection, or a public Wi-Fi hotspot.

By contrast, when you configure your web browser to route web traffic through a proxy server, then access to the otherwise blocked website is unrestricted. Why? Because from the perspective of your ISP or Internet access provider, you are connecting to a "safe" server. The proxy server masks your true web browsing targets from your ISP.

Now for the good news: You can configure your Raspberry Pi as both a VPN endpoint as well as a proxy server! Take a look at Figure 17.2.

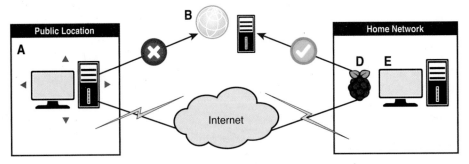

FIGURE 17.2   Network topology for a Raspberry Pi-based VPN and proxy solution

Let's step through each point in this process:

A. This is your laptop, smartphone, or other Internet-connected device that connects to the Internet over a public, unsecure medium. This is most likely a public Wi-Fi hotspot.

B. Let's imagine that this icon represents your target website. Your concern may be that you don't want your logon credentials sent over your network connection in an unsecure fashion, or maybe you simply cannot connect to the target site because your Internet access provider blocks it.

C. The solution to this problem is connecting your Internet-connected device to both your configured Hamachi VPN, of which your Raspberry Pi is a member; as well as your Raspberry Pi proxy server. (I don't show C in Figure 17.2.)

D. The Raspberry Pi, because it is a member of your internal LAN as well as your Hamachi VPN, gives you remote access to internal network resources (E) in a completely protected manner. The Pi also cloaks your point of origin on the Internet because from the perspective of your public Internet access provider, you're not connecting to your target website, but instead to your Raspberry Pi in your home location.

Building Your Raspberry Pi VPN Gateway

Let's get this party started, shall we? First, some good news: The good folks at LogMeIn offer Hamachi VPN for free for up to five hosts. For this test configuration, you need just two members: one is your Raspberry Pi, and the other is your desktop computer.

Before you go any further, go to the Hamachi website and register a free user account: http://is.gd/Njxokw. Your credentials will consist of an email address and a password.

NOTE: SECURE REMOTE ACCESS

In this chapter I use Hamachi as a means to an end, in other words, as a secure platform for web proxy services. However, you should be aware that Hamachi also gives you an excellent way to connect to your home network securely from anywhere in the world. Remember that all the VPN traffic is tunneled over standard web ports, so you don't have to worry about firewall exceptions or other Internet access filters. LogMeIn has some awesome technologies!

Next, you need to download and install the Hamachi client on your desktop workstation. Visit http://is.gd/ruIvfl and get the software; LogMeIn has all the major platforms covered:

- Windows
- OS X
- iOS
- Android

You can use the Hamachi desktop client to create VPNs, but you get much more flexibility by doing so from the web portal.

TASK: CREATING YOUR HAMACHI VPN

Let's get Hamachi up and running, shall we? You don't need to complete this set of tasks from your Raspberry Pi, necessarily. Any computer or even mobile device that is Internet-connected and has a standard web browser is fine.

1. Visit https://secure.logmein.com and log in with your account email address and your password.

2. From the left-hand navigation menu in the management console, click Networks.

3. In the Networks area, click Add Network. Add a name and optional description for your new VPN. As you see in Figure 17.3, Hamachi supports three different network topologies, each with its own characteristics.

I suggest that unless you have a compelling reason to do so that you select the Mesh network type. This configuration allows you to connect to and interact with all devices on your home or personal network. For additional details, LogMeIn publishes a wonderful Hamachi user guide; the document is available for free download at http://is.gd/tbNsMs.

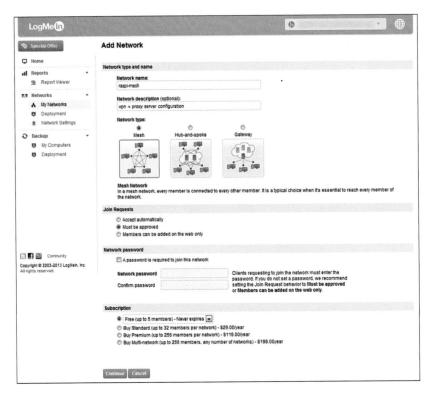

FIGURE 17.3   It's simple to create private, encrypted networks by using Hamachi.

4. There are two pages to the Add Network screen (though I've combined them in Figure 17.3). Click Continue when you're ready to proceed to the second page.

5. The final step in creating your new network is specifying security options. I suggest you leave the default, which requires that you (the administrator) approve any requests to join your new VPN. You can add a password as well to increase security even further.

6. Finally, leave the Free subscription level as is and click Continue to proceed.

When you view the My Networks node in the LogMeIn web console, you see all details concerning your new private network.

Now that you've reserved a VPN in the Hamachi infrastructure, let's get your desktop client connected to it.

TASK: CONNECTING TO YOUR HAMACHI VPN FROM YOUR DESKTOP COMPUTER

"The proof is in the pudding," said my old mentor, Bernie Carr. We can't demonstrate how VPN technology works until we actually, well, demonstrate the technology. Let's get to work!

1. Start up your Hamachi client and click the Power On button. I show you the interface screens for my Windows 7 computer in Figure 17.4.

2. Provide a name (Hamachi ID) for your client computer. Make it meaningful so you'll know instantly which computer is which when you view your VPN.

3. You now see your computer's name and an IP address in the 25.x.y.z range in your Hamachi Control Panel. Click Join an existing network and then provide the nine-digit Network ID and optional password. You can fetch the Network ID from the web portal when you click your VPN. Next, click Join to join the network.

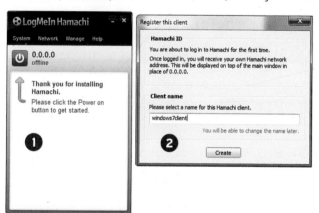

FIGURE 17.4 The LogMeIn Hamachi client looks and behaves the same way on Windows and OS X.

If you created your Hamachi VPN with the defaults, you get a message informing you that the network requires manual approval for new members. Click Yes to submit a request for membership.

4. Log in to the web portal, where you see a join request for the desktop client. Accept the request and click Save to approve. You see the Hamachi desktop client update immediately.

You can now control access to the VPN by right-clicking the network name in the Hamachi client and selecting either Go Offline or Go Online as the case may be.

I help you work with your new configuration a bit later in the chapter. It's time to install the Hamachi client on your Raspberry Pi.

TASK: INSTALLING HAMACHI CLIENT ON YOUR RASPBERRY PI

1. Establish an SSH session to your Raspberry Pi. To allow you to execute root commands without having to type sudo before every command, run the Bash shell as root:

```
sudo bash
```

2. The Hamachi client for Linux has a prerequisite: the Linux Standard Base (LSB) core libraries. Install them in the usual manner:

```
apt-get install -y --fix-missing lsb-core
```

3. Install the Hamachi Client for Linux. You should substitute the package I supply in the sample syntax with the latest version on the Hamachi Labs home page at http://is.gd/Lays35. Note that LogMeIn provides Intel 32-bit, Intel 64-bit, and ARM versions of the software. You *do* remember that the Raspberry Pi is an ARM device, correct?

```
wget https://secure.logmein.com/labs/logmein-hamachi-2.1.0.101-armel.tgz
tar -zxvf logmein-hamachi-2.1.0.101-armel.tgz
cd logmein-hamachi-2.1.0.101-armel
./install.sh
```

4. Start the Hamachi service (daemon):

```
/etc/init.d/logmein-hamachi start
```

5. The next three commands perform the following actions:

- Logging the Pi into the Hamachi network
- Binding the Pi to your LogMeIn account
- Specifying an ID for the Pi

```
hamachi login
hamachi attach <your_email_address>
hamachi set-nick raspi-proxy
```

6. From your desktop workstation, approve the network join request from the web console. You can see the interface in Figure 17.5.

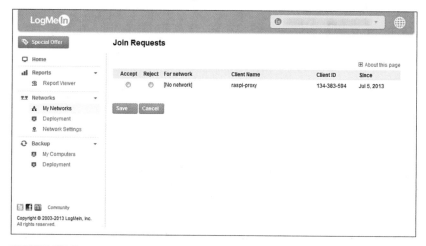

FIGURE 17.5   Approving a Hamachi join request

7. Navigate to the My Networks page and add your Raspberry Pi to the Hamachi VPN you created earlier. Note that approving a connection request and joining a particular network are two separate steps.

You now can communicate directly with your Raspberry Pi by connecting to the Pi's 25.x.y.z IP address from another Hamachi network member anywhere in the world!

Hamachi's ability to allow easy peer-to-peer networks is one reason why the service is so popular among gamers. Many first-person-shooter and Minecraft players, for instance, leverage Hamachi to allow for easy LAN gaming sessions.

Building Your Raspberry Pi Proxy Server

As I said earlier, a proxy server is a computer that stands in place of another computer in terms of making web requests. Actually, that's what the noun proxy means in the first place.

Besides providing confidentiality to hosts located behind the proxy server, the other benefit of the proxy service is that of caching. In other words, proxy servers can boost your browsing speed by serving up web pages cached locally instead of having to fetch the content from the source every time.

Nowadays you don't need specialized hardware and software to set up a proxy server. In this chapter we use the free and open source Privoxy (pronounced *prih-VOX-ee* from http://privoxy.org). Privoxy is fast, easy to configure, and flexible. However, you should be aware that Privoxy is a non-caching HTTP proxy, which is fine because the goal here is confidential, restriction-free web browsing, not necessarily a performance boost.

Before you undertake the following procedure, make a note of the 25.x.y.z IP address that Hamachi reserved for your Raspberry Pi. You use that IP address as the proxy server endpoint address.

NOTE: WHAT DOES 25.X.Y.Z MEAN?

As it happens, LogMeIn owns at least a portion of the 25.0.0.0/8 IPv4 address space. Thus, the Hamachi service can dish out globally unique IP addresses to its customers within this range. You might recall that all hosts on the same IP subnet can communicate directly with each other.

TASK: INSTALLING AND CONFIGURING PRIVOXY RASPBERRY PI

In this procedure, you get Privoxy up and running on your Raspberry Pi. As you would expect, you should perform the following steps from a Terminal session on your Pi.

1. Establish an SSH session to your Raspberry Pi and put your shell session in the root user context.

```
sudo bash
```

2. Download and install Privoxy from your default software repositories:

```
apt-get install -y privoxy
```

3. You need to make a couple tweaks to the Privoxy configuration file.

```
nano /etc/privoxy/config
```

4. Use the nano keyboard shortcut Ctrl+W to search for the string listen-address. When you find it (you'll have to scroll past a few screens of documentation before you get to the uncommented value), edit the line like so:

```
listen-address 25.x.y.z:8118
```

5. Substitute your Raspberry Pi's actual Hamachi IP address for the example given here.

6. Save your work, close the file, and restart the Privoxy service.

```
service privoxy restart
```

Testing the Configuration

Alrighty then! You have your VPN and proxy all set up and ready to test. Sit down at your desktop computer and turn on your Hamachi network. Verify that your Raspberry Pi also shows up in your Hamachi Control Panel.

Open your web browser and navigate to the following website:

```
http://privoxy.org/config
```

The resulting configuration page should say *Privoxy is not being used*. This is an expected result because you haven't configured your desktop PC to route HTTP web traffic through your Raspberry Pi proxy. You can see what this page looks like in Figure 17.6.

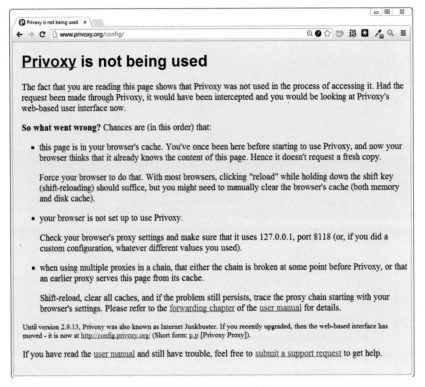

FIGURE 17.6 The Privoxy config page tells you instantly whether or not your browser session is being proxied.

Now point your browser to the IP Info Database (http://is.gd/23sxDf) or an equivalent site and verify your system's public IP address and geolocation. Of course, you need to do this from an IP address/location different from where your Raspberry Pi is located.

Fortunately, I have some computers located in different areas of the United States. To that point, I show you the before proxy information for my Windows 7 workstation, located in Columbus, Ohio, in Figure 17.7.

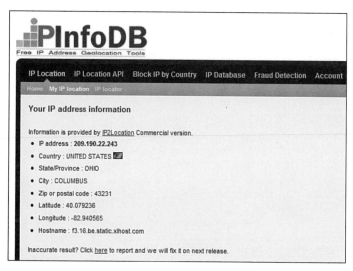

FIGURE 17.7 My "pre Proxy" IP address has my workstation located in Columbus, Ohio.

Now you need to configure your web browser to use a proxy server. For this example I've used Internet Explorer, but here are some references to online tutorials to perform this action on other popular browsers:

- **Chrome (Windows, OS X)**: http://is.gd/RB2WYa
- **Chrome (Android)**: http://is.gd/XRoYeS
- **Firefox**: http://is.gd/oTyHmD
- **Safari (OS X)**: http://is.gd/9cZgje
- **Safari (iOS)**: http://is.gd/znmAwg
- **Opera**: http://is.gd/XI3KJ9

TASK: POINTING YOUR WORKSTATION COMPUTER AT YOUR PI PROXY

This task assumes that we are working from a Windows 7 or Windows 8 computer.

1. Open up the Internet Explorer web browser and open the gear menu in the upper right of the window. Next, click Internet Options.

2. In the Internet Properties dialog box, open the Connections tab and then find and click the LAN Settings button. I show you both dialog boxes in Figure 17.8.

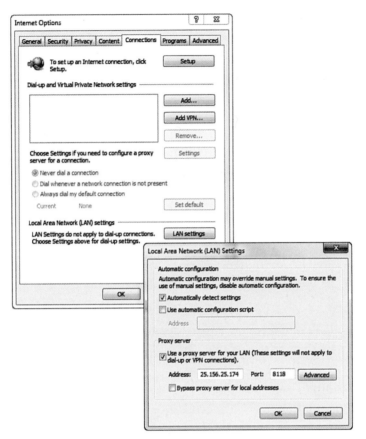

FIGURE 17.8   The process of configuring a web browser to tunnel traffic through a proxy is about the same regardless of the application or OS platform.

3. Under Proxy Server, select the option Use a proxy server for your LAN and enter your Raspberry Pi's Hamachi IP address. Make sure to use 8118 as the port number.

4. Click OK out of all dialog boxes and restart your browser.

Go back to the IP Info DB website and recheck your IP address and location. You should find that the site reports your IP address and geolocation as that of your home network (where your Raspberry Pi proxy server is located, in other words). I show you this in Figure 17.9.

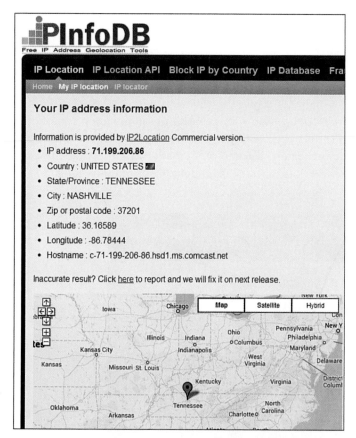

FIGURE 17.9 My workstation's "post Proxy" address shows it as originating in Nashville, Tennessee, where my Raspberry Pi resides.

For the duration of your web browsing session, you not only mask your workstation's IP address location, but you also provide for data confidentiality because all traffic flowing within the Hamachi VPN is fully encrypted.

Just for grins, I figured that you would be interested in the "What's My IP" type websites. To that end, here are a few more for you to try out and experiment with (some expose an API that is scriptable!):

- **What is My IP?**: http://is.gd/uIhmYK
- **IP Chicken**: http://is.gd/WRPm0J
- **WTF is My IP?**: http://is.gd/a6ARdP
- **IP2Location**: http://is.gd/S1cd19

Speaking of Geolocation...

Do you remember in Chapter 4, "Installing and Configuring an Operating System," when I briefly mentioned the Rastrack website? Let's spend just a couple minutes discussing it now, as its operation relates somewhat to privacy and security issues.

Rastrack (http://is.gd/sGStJL) is a map that shows the general location of Raspberry Pis throughout the world. Of course, only Pi owners who volunteer this information contribute to the map. Also Ryan Walmsley (http://is.gd/lUb70X), a British high-school student who created the site, has no verification scheme in place to prove that registrations come from actual Raspberry Pis. You can see what the Rastrack map looks like as of this writing in Figure 17.10.

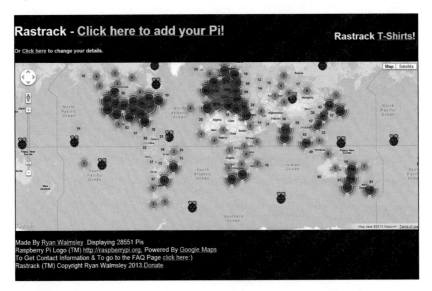

FIGURE 17.10 The Rastrack map, as of summer 2013

Given the context of this chapter, you might wonder if Rastrack uses IP address geolocation to determine the physical location of your Pi. The answer is no and yes.

Ryan has not specifically coded any geolocation into Rastrack. Instead, he plots user-provided data through the Google Maps API, which gives a general location based on a combination of your ISP's IP addressing metadata and the address information you provide to the site.

For instance, check out Figure 17.11. My Zip code 37221, and the map result places my Pi directly in the middle of that Zip code zone, not necessarily anywhere near the Pi's actual location on the globe.

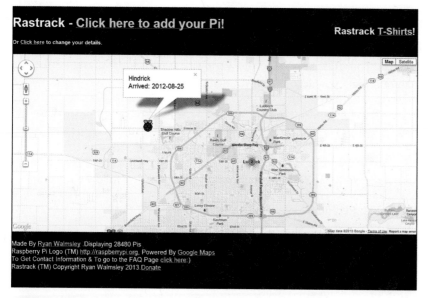

FIGURE 17.11 The Rastrack map plots location based on user-specified information and Google Maps metadata.

Why would you want to add your Raspberry Pi to Rastrack? Well, I've found that the Raspberry Pi community is a passionate one and folks are eager to share information with each other. It's pretty cool to see all the Raspberry Pi devices registered in most countries in the world.

By contrast, any time you expose computer system details to the public, that opens the door for a malicious user to think, "Aha! There are four Raspberry Pis in my neighborhood. Maybe I'll try some reconnaissance and see if I can penetrate their networks!" Sad to say, this kind of activity is much more common than you might think.

CAN I HAZ COPEY EDITR?

I'm sure it is just the "author" in me, but I was appalled by the many egregious spelling and grammatical errors present on the Rastrack site—and the Add Your Pi! pages in particular. I had to remind myself that this web app was created by a high school student. At any rate, perhaps Ryan will correct these mistakes by the time you access the site.

TASK: ADDING YOUR RASPBERRY PI TO THE RASTRACK DATABASE

Should you decide to add your Raspberry Pi to the worldwide Rastrack database, I offer you the following procedure to help you accomplish your goal.

1. From LXDE on your Pi, open up Midori or your favorite web browser and browse to the Rastrack website:

```
http://rastrack.co.uk/
```

2. Click Click here to add your Pi and fill out the form by providing the following details:

- **Name or Nickname**: Required.
- **Twitter Username**: Optional.
- **Date of Arrival**: Required. Use format YYYY-MM-DD.
- **Email**: Required. This field is important because it serves as your ID if you ever want to edit or remove your Pi location registration.
- **Location**: Required. I had the best luck by providing a postal (Zip) code here.
- **Human Verification**: Required. This field is meant to prevent spam bots from submitting erroneous registrations. The technology used here and in many websites is called Completely Automated Public Turing Test to tell Computers and Humans Apart (CAPTCHA, and no, I'm not kidding). You can learn more about CAPTCHA by visiting http://is.gd/zsKPcW.

If you need to change your registration details or remove your listing, visit the Rastrack site and click the Click here to change your details link. You are asked to provide the email address you used when you originally registered your Pi. You'll be sent an email message with a key and link to change your registration details.

Building a Raspberry Pi Tor Proxy

Have you heard of The Onion Router (Tor) network? Tor (http://is.gd/tQul4e) is free software as well as an open network that provides users with excellent privacy by routing network traffic through a series of distributed Tor routers, none of which has knowledge of the complete end-to-end path of the communications.

Take a look at Figure 17.12, and I'll walk you through the basic mechanics of Tor.

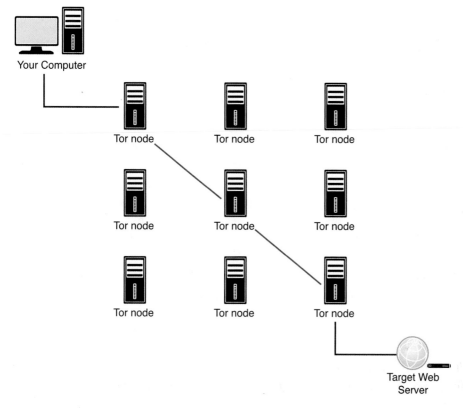

FIGURE 17.12 Schematic of the Tor anonymity network.

The client software running on your computer selects a random path through the Tor network for every data packet that is transmitted. The intermediate Tor routers, which are formally called *nodes*, have no knowledge of the full routing path; they basically forward each packet to a randomly selected next hop address and forget about the transmission.

As I said earlier, the exit node is the vulnerable point in the Tor network because it is possible for a malicious individual to configure his computer as a Tor exit node, sniff unencrypted traffic from the Tor network, and glean potentially privacy-busting data about the data transmission originator.

The good news is that you have to take deliberate configuration steps to become a Tor exit node. For the vast majority of Tor users, their data is safe so long as they practice good web browsing hygiene such as enabling SSL and not leaking any personal information via web forms.

Routing select web traffic through the Tor network gives you the following advantages:

- Your source IP address and geolocation are completely obfuscated to anybody who tries to execute a man-in-the-middle attack on your computer.

- All data transmitted within the Tor network is encrypted. However, unless you are using HTTPS or another encryption technology, your unencrypted data that enters the Tor network emerges from that network equally unencrypted.

The main disadvantage to Tor is the same as what we see with some corporate VPNs; namely, tremendously slow speeds. You do not want to use Tor for ordinary web browsing, trust me. Network traffic through the Tor network moves slower than I remember browsing with a 14.4Kbps analog modem in the mid-1990s. Don't even think of sending or receiving binary files over the Tor connection. Doing so is disrespectful to the people who make up the Tor community. Instead, use the Tor network only when privacy is your principal concern.

In this section I want to give you the high-level overview for configuring your Raspberry Pi as a Tor proxy. Due to space constraints (the actual step-by-step is fairly involved), I'm going to turn you over to Adafruit, which worked up a couple wonderful tutorials on how to enact this configuration:

- Adafruit Raspberry Pi Wireless Access Point Tutorial: http://is.gd/MlMNEP
- Onion Pi Tutorial: http://is.gd/7EHgqx

NOTE: WHAT DOES AN ONION HAVE TO DO WITH AN ANONYMOUS NETWORK?

The onion is not only the logo icon for the Tor project, but also represents the network itself and a non-public DNS top-level domain. As it happens, you can build your own web server that exists entirely within the Tor (onion) network. These so-called hidden service sites use special DNS addresses called onion URLs that end with the .onion domain.

You can view a schematic of the Raspberry Pi Tor proxy in Figure 17.13.

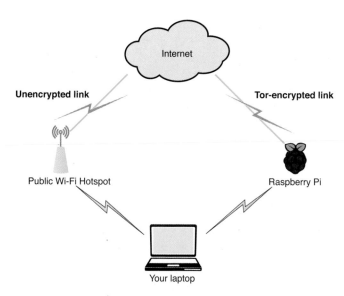

FIGURE 17.13   Schematic showing how you can set up your Raspberry Pi Tor proxy and wireless access point

As you can see, you can install open source software to configure your Raspberry Pi as a wireless access point. Because the Pi is multi-homed (that is to say, it has more than one network interface), you can use the Wi-Fi connection as your entry point for Tor communications and the wired Ethernet connection as your gateway to the Internet.

Therefore, the workflow for using this special Pi is as follows:

- You bring your Onion Pi device to a hotel, conference, or other location and plug the Ethernet interface into a live Internet connection.

- You then use your laptop or mobile device to join your personal Onion Pi Wi-Fi network whenever you need Tor-protected browsing.

## CHAPTER 18

Raspberry Pi Overclocking

If you've made it this far into the book, then you've probably given your Raspberry Pi quite a workout. Accordingly, you are well aware of the hardware limitations of the diminutive Model A or Model B board.

How can you squeeze more performance out of your Pi? Can you do so safely? How does overclocking affect the limited warranty offered by Farnell and RS Components?

In this chapter I begin with a comprehensive tutorial on how to overclock your Raspberry Pi. Of course, you first need to understand exactly what overclocking means. Next, I give you everything you need to know to improve the performance of your Pi while staying within the bounds of warranty. Of course, if you want to go outside the bounds, then that's okay, too. Just don't knock on my door when something goes wrong!

I also address how to adjust the split between CPU and GPU memory allocation. For instance, if you work from the shell prompt, there is no reason to allocate any more RAM than is absolutely necessary to the GPU. On the other hand, if you have your Pi set up as a RaspBMC media center, then just the opposite may be true.

Let's get to work!

What Is Overclocking?

Overclocking refers to forcing a computer component, such as the central processing unit (CPU), to operate faster than its default clock frequency.

The Pi's ARM ARM1176JZF-S processor operates with the following default frequencies:

- **CPU**: 700 million clock cycles/second (MHz)
- **GPU**: 250MHz
- **SDRAM**: 400Mhz

The Raspberry Pi is capable of processing one command per clock cycle, which means that the CPU processes 700 million instructions per second, the GPU processes 250 million, and the RAM chip 400 million. Those are a lot of instructions!

If you can force the Raspberry Pi to increase the clock rate for the CPU, GPU, or RAM, then it follows logically that the Pi will run faster by executing more commands per unit time. True enough.

The downside to the overclocking situation is heat. Hopefully it makes sense to you that an overclocked SoC requires more power and therefore generates more heat than an SoC running at its defaults.

You remember that the ordinary operating voltage of the Pi is 5V and that the typical current draw is between 700 and 1400 mA.

NOTE: POWER TO THE PI

All this talk of overclocking and performance improvements assumes that you provide the Pi with steady, reliable power. To do that you need to make sure you use a quality power supply before attempting to overclock your board.

Although overclocking in itself is reasonably safe, overvolting on the other hand will likely reduce the lifetime of your Pi by gradually degrading the SoC's transistors and logic gates.

Some enthusiasts, myself included, aren't overly worried about slightly reducing the lifetime of our Pis because the cost of replacement is eminently reasonable, but it's a factor you should consider nonetheless.

Warranty Implications of Overclocking

The following is a relevant extract from Farnell's Raspberry Pi Limited Warranty (http://is.gd/jF9ELL):

> What does this limited warranty NOT cover?
>
> Newark element14 has no obligation to repair, replace, or provide refunds in the following instances:
>
> If the alleged defect arises because Customer has altered or repaired the Raspberry Pi without the prior written consent or authorization of Newark element 14

As I get to shortly, the Raspberry Pi Foundation does offer overclocking modes that do not void the warranty. Using these modes still qualifies you for a refund from the distributor.

The salient question on the mind of the enthusiast, is "How would Farnell or RS Components know if I tried to overclock or overvolt my Pi?"

As it happens, the Foundation programmed the SoC such that a so-called sticky bit is turned on when any of the following conditions is detected:

- You set the temperature limit to a value above 85 degrees Celcius.
- You force Turbo Mode or disable the current limit and set an overvoltage.

You can get a detailed description of exactly which overclocking options void the warranty on the eLinux.org website: http://is.gd/1HcNWb.

More about the sticky bit: As I said, this is a flip flop circuit that, once tripped, remains in place and is used by the distributors to detect whether you set your Pi to warranty-breaking settings.

You can tell if your Pi's sticky bit is set by running the following command from a shell prompt and reviewing the Revision value:

```
cat /proc/cpuinfo
```

The word on the street is that any Revision value above 1000 means that your sticky bit is set and you do not qualify for a refund from the distributor.

Take a look at Figure 18.1. The first output shows my original Pi settings, which fall within the bounds of the warranty spec. The second output shows that the sticky bit has been tripped as a result of my setting an illegal overvoltage value.

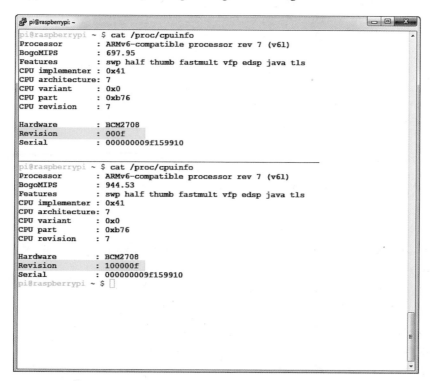

FIGURE 18.1   This output shows my Raspberry Pi both pre and post-sticky bit.

I verified that resetting the run status of my Pi did nothing to change my CPU Revision value. To the best of my knowledge, after the sticky bit is flagged, your Pi is permanently out of warranty.

Overclocking Your Pi

Now that you understand what's involved in overclocking, are you still willing to dig in, get your hands dirty, and boost the performance of your Pi? Great...so am I.

The Raspi-Config Method

The easy way to overclock your Raspberry Pi and to stay within warranty boundaries is to use the Raspi-Config script.

Start Raspi-Config by running the command sudo raspi-config and select the Overclock option from the main menu. Press Enter to go past the warning screen. You see that you can select one of five overclock presets; these are summarized for you in Table 18.1.

TABLE 18.1   Raspberry Pi Overclock Modes

Preset	ARM Freq	CPU (core) Freq	SDRAM Freq	Overvolt
None	700MHz	250MHz	400MHz	0
Modest	800MHz	250MHz	400MHz	0
Medium	900MHz	250MHz	450MHz	2
High	950MHz	250MHz	450MHz	6
Turbo	1000MHz	500MHz	600MHz	6

The ARM/GPU core voltage values are a trifle mysterious. The default value of 0 denotes 1.2V, and a value of 6 represents 1.35V. The voltage increases in 0.024V steps between 0 and 6. You can overvolt up to a value of 8, which is 1.4V.

After you select an overclock preset, you're prompted to reboot your Raspberry Pi for the change to go into effect.

The Raspberry Pi Foundation is rightly proud of the Turbo mode option; to that point, I'd like to explain how it works in more detail.

In addition to providing a big performance boost to your Pi, the Turbo mode dynamically adjusts your CPU, GPU, and SDRAM frequencies depending on the load your Pi experiences.

When the SoC temperature reaches its threshold value, Turbo mode automatically scales back the Pi speed to allow the system to cool down. Pretty cool, eh?

This Turbo mode magic is the result of a Linux kernel driver called cpufreq, which serves as a governor, or controller, over the Pi's overclock status.

A Swedish programmer named Enrico Campidoglio wrote a cool Bash shell script that gives you your Pi's CPU status details. The script also gives you "the deets" regarding voltage and temperature.

TASK: VERIFYING YOUR PI'S CPU, VOLTAGE, AND TEMPERATURE STATUS

This procedure should be accomplished from a terminal session (local or remote) on your Raspberry Pi.

1. Copy Enrico's shell script contents (http://is.gd/h4q135) and paste the data into a new, blank file on your Pi. Save the file as cpustatus.sh.

2. Open a shell prompt and run the script from a Terminal session. You first have to mark the file as executable, however. For instance, the following shows you my command statement assuming that the script exists in my present working directory:

```
chmod +x cpustatus.sh
```

3. Cool! Now that the shell script can be run as executable code, let's go ahead and run the script.

```
./cpustatus.sh
```

Remember that in Linux, you use the "dot slash" (./) notation to tell Raspbian that you want to run the given executable program from the present working directory. Otherwise, you would have to supply the full path to the file, such as /home/pi/cpustatus.sh. Sample output from the script file on my Raspberry Pi Model B is shown in Figure 18.2.

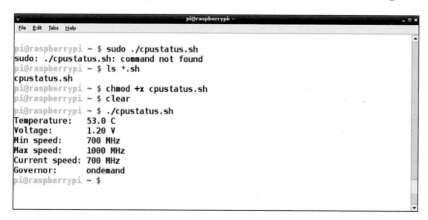

FIGURE 18.2 This Bash shell script provides you with useful, at-a-glance CPU metadata.

The Manual Method

The power user way to overclock or overvolt your Raspberry Pi is to manually edit the /boot/config.txt file. Recall that the Raspi-Config script, which is itself located in usr/bin, is nothing but a user-friendly front-end to the config.txt file.

From a shell prompt, let's open the file for editing:

```
sudo nano /boot/config.txt
```

Scroll to the end of the file to see the parameters that are relevant to this discussion. Here are the default entries:

- **arm_freq**: CPU clock frequency
- **core_freq**: GPU clock frequency
- **sdram_freq**: RAM clock frequency
- **over_voltage**: Degree of overvoltage

The good news is that you can include additional parameters to take full control over our Pi's overclocking experience. Here are three juicy options, as documented at the eLinux.org website (http://is.gd/1HcNWb):

- **temp_limit**: You can modify the overheat protection by adjusting this threshold value. The default value is 85 degrees Celsius.
- **current_limit_override**: Disables current limit protection. Remember that electrical current is directly proportional to voltage; enabling this option can help if your Raspberry Pi suffers reboot failures due to your configuring the overclock setting too high.
- **force_turbo**: Disables the cpufreq driver and sets the Pi to run with highest settings all the time.

NOTE: HELP! MY PI WON'T BOOT

If during your overclocking and overvolting experiments you find that your Pi refuses to boot, then don't fret. Instead, unplug the Pi, wait a couple of minutes, and then power on the device while holding down the Shift key. This disables any overclocking settings for the current boot only. You can then reset your Pi back to safer values, reboot again, and you should be good to go.

Benchmarking Your Pi

Okay...you are probably thinking, "Tim, I understand everything you've taught me so far, and I've overclocked my Pi. Although the device *feels* like it's running faster, how can I prove this quantitatively?"

I'm glad you asked! *Benchmarking* refers to running tests that compare the current values of a process either to past values or to values generated by other, related processes.

Specifically with reference to the Pi, you can run benchmark tests against the device itself to gauge performance changes, such as before and after overclocking the processor cores and RAM.

There are a number of benchmarking utilities for Linux in general and the Raspberry Pi in particular. Here is a run-down of some of the most popular utilities and their associated websites in no particular order:

- **HardInfo**: http://is.gd/sW7i7D
- **GtkPerf**: http://is.gd/lsKamm
- **nbench**: http://is.gd/Ql0MXe
- **Quake III Arena timedemo**: http://is.gd/SBkoQq (building Quake III in Raspbian); http://is.gd/1p553T (timedemo instructions)

Because it is the benchmark used by Eben on the Raspberry Pi website, you learn here how to use nbench.

Interestingly, nbench (http://is.gd/mlJ2FO) is a command-line computer benchmarking utility that was developed originally by the long-deceased *BYTE* magazine in (wait for it) the mid-1990s. The tool measures a computer's CPU, floating-point math co-processor, and memory subsystems by comparing your system's results to two archaic reference machines:

- Dell Pentium 90MHz with 256KB cache RAM and running MS-DOS
- AMD/K6 233MHz with 512KB cache RAM and running Linux

The nbench software runs 10 tasks that each analyze a specific component of your target system's performance; you can read more detail about the tests and algorithms by visiting http://is.gd/clvnY8. The system data nbench amasses is then compared to stored baseline data for the 90MHz and/or the 233MHz machines. The nbench in no way emulates the reference machines; all it has is stored benchmark data.

TASK: INSTALLING AND RUNNING NBENCH

Even after all these years, nbench can be a worthwhile performance benchmarking tool. Let's run the tool against our Raspberry Pi.

1. Open a shell prompt and begin by downloading the nbench source code, unpacking it, and compiling it.

```
wget http://www.tux.org/~mayer/linux/nbench-byte-2.2.3.tar.gz
tar xzf nbench-byte-2.2.3.tar.gz
cd nbench-byte-2.2.3
make
```

The make step is doubtless unfamiliar to non-Linux computer users. To be sure, it's a different paradigm to think of obtaining a program's source code and compiling it into executable form manually. Thus, the make program assembles all the parts and pieces that comprise a piece of software and compiles the code into executable form.

2. Make sure you have everything but your Terminal session closed or stopped. Next, start the nbench tool.

```
./nbench
```

In Figure 18.3 you can see my benchmark results using my Model B board that runs in Turbo mode.

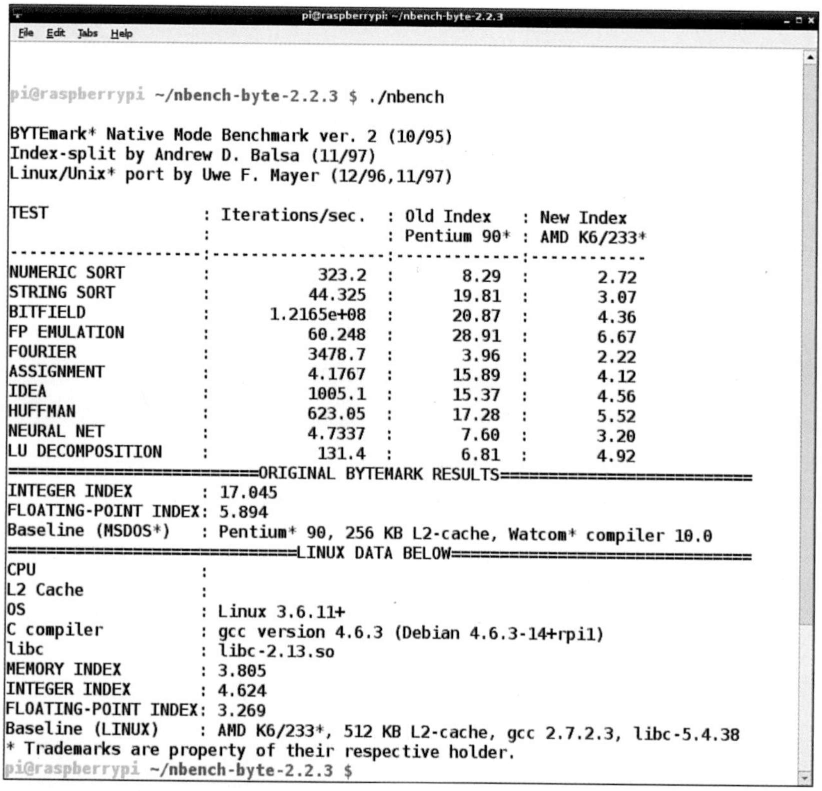

FIGURE 18.3 The BYTEmark (nbench) benchmark is particularly popular among Raspberry Pi enthusiasts.

It's one thing to run a benchmark program and quite another to understand what it means. You can see that nbench puts your Raspberry Pi through 10 different tasks. What's important, speaking analytically, isn't comparing your system's results to the reference systems. Instead, you should compare the figures against your own Pi say, before and after enabling overclock settings. You can also compare your nbench results with other Raspberry Pi users, for instance, on the Raspberry Pi Forums (http://is.gd/6nBR5Z).

Adjusting the Memory Split on Your Pi

In addition to overclocking, you also have the ability to modify the CPU/GPU memory split to accentuate system performance on your Raspberry Pi.

You already know that the Broadcom BCM 2835 SoC has two processors (CPU and GPU), as well as the SDRAM. You also know that the Model B board includes 512MB of RAM, while the Model B board includes 256MB.

You can adjust the balance between how much RAM is allocated to the CPU vs. the GPU. This can be helpful if, for instance, you perform mostly graphics-intensive tasks such as playing games or transcoding video.

On the other hand, if your Pi is set up to run only or mostly with the Terminal, then it does not make sense to allocate any more RAM than necessary to the VideoCore IV GPU.

As you read with the overclocking discussion, you can tweak the memory split either with Raspi-Config or manually.

TASK: TWEAKING MEMORY SPLIT WITH RASPI-CONFIG

In this procedure you take control over the CPU/GPU memory split on your Raspberry Pi. Perform these steps on your Pi.

1. Open Raspi-Config and navigate to Advanced Options. Next, select Memory Split.

2. You are asked the question *How much memory should the GPU have?* Supply a legal value and press Tab to select OK. The legal values for the Model B are the following, which represent megabytes (MB) of RAM:

- 16
- 32
- 64
- 128
- 256

NOTE: REGARDING THE RAM SPLIT ON THE RASPBERRY PI MODEL A

The RAM allocation values presented here are obviously for the Model B board. The Model A board supports all values less than or equal to 128.

3. You are prompted to reboot, after which your Pi reserves the designated amount of RAM to the GPU. Any RAM left over is left by default to the CPU.

As far as suggested splits are concerned, I share with you here what I use on my Model B boards. For servers that aren't doing much of anything, I use a 240/16 CPU/GPU split. For my gaming rigs and media center devices, I use a 256/256 split with no issues.

Tweaking Memory Split by Hand

We learned in Chapter 17, "Raspberry Pi Security and Privacy Device," what /boot/config.txt is and how to edit this crucial system configuration file. Thus, we get directly to the relevant parameters:

- **gpu_mem**: RAM devoted to the GPU. The CPU gets any remaining memory.
- **cma_lwm**: When the GPU has less than this low water mark amount of RAM, it requests more from the ARM CPU.
- **cma_hwm**: When the GPU has more than this high water mark amount of RAM, it releases some to the CPU.

To set the cma_lwn or cma_hwm dynamic memory split parameters, you need to add the following line to your /boot/cmdline.txt file:

```
coherent_pool=6M smsc95xx. turbo_mode=N
```

NOTE: FOR THE ESPECIALLY BRAVE...

Before experimenting with the dynamic memory split parameters, I advise you to update your Linux kernel to the latest available. I gave instructions for doing so back in Chapter 4, "Installing and Configuring an Operating System."

The cmdline.txt configuration file contains low-level commands that are sent directly to the Raspbian Linux firmware at boot time. The file consists of key/value pairs separated by a space. Therefore, you should just add the line of code just given as a new entry to the file, save changes, and reboot.

To see your Pi's current cmdline data, run the following command from a shell prompt:

```
cat /proc/cmdline
```

A Historical Footnote

In earlier Raspbian releases, the /boot partition contained multiple GPU firmware files:

- start.elf
- arm128_start.elf
- arm192_start.elf
- arm224_start.elf
- arm240_start.elf

The original idea was that you could change the CPU/GPU memory split by overwriting the "live" start.elf file with one of the arm* files. Thus, by running the following example statement:

```
cp /boot/arm240_start.elf /boot/start.elf
```

you configure your Pi with a 240MB ARM/15MB GPU split. Nowadays the /boot partition contains just a single start.elf GPU firmware image. You can see the Raspberry Pi kernel, firmware, and configuration files in Figure 18.4.

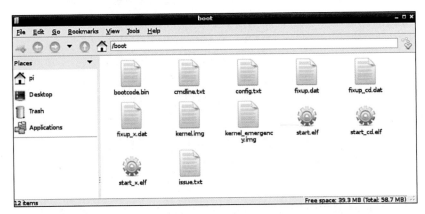

FIGURE 18.4 The Raspberry Pi kernel, firmware, and boot files are located on the FAT32 /boot partition (the only part of the SD card that is readable by Windows).

Just for the sake of completeness, I want to provide you with a brief description of the /boot partition contents; check out Table 18.2 for that useful nugget of information.

TABLE 18.2 Raspberry Pi /boot Partition Contents

File Name	Description
bootcode.bin	The second-stage bootloader
cmdline.txt	Parameters that are passed to the kernel
config.txt	Configuration file that is read by the GPU
fixup.dat	Configures the SDRAM partition between the GPU and the CPU
fixup_cd.dat; fixup_x.dat	Testing versions of fixup.dat (do not use)
issue.txt	Gives you the manufacture date of your Pi (run cat /boot/issue.txt)
kernel.img	The ARM operating system kernel
kernel_emergency.img	Enables you to repair the Linux partition using e2fsck if it ever becomes corrupted
start.elf	GPU firmware image
start_x.elf; start_cd.elf	Alternate versions of start.elf that are invoked in certain memory split situations

Remember our friend Hexxeh (http://is.gd/Rrh2bS)? This is the individual who gave us a splendid, easy-to-implement method for updating Raspberry Pi's firmware:

```
sudo apt-get install rpi-update
sudo apt-get install git-core
sudo wget https://raw.github.com/Hexxeh/rpi-update/master/rpi-update -O /usr/bin/
rpi-update && sudo chmod +x /usr/bin/rpi-update
sudo rpi-update
```

NOTE: WHAT IS GIT?

You probably noticed that many of the installation recipes in this book use a software product called Git. Git (http://is.gd/5vKXkJ) is a free, open source, distributed software version control system. Git enables developers to publish their projects and source code so that people can access their stuff in myriad ways. If you have any interest in writing your own open source software, you want to spend time getting to know Git.

The reason I mention Hexxeh's tool is that it once allowed you to change the ARM/GPU memory split. Now, however, Hexxeh recommends using the Raspi-config or manual methods as previously outlined.

Various and Sundry Performance Tweaks

The IT security and performance principle of *least service* means that if your computer does not need to have a particular background service (daemon) running, then prevent it from doing so. Not only will your computer run faster, but you also reduce the attack surface of the machine. After all, an attacker can't compromise a service that isn't running!

To that end, you can use the sysv-rc-conf utility (http://is.gd/1VjHLP) to analyze your startup services and disable any that you identify as unnecessary.

```
sudo apt-get install -y sysv-rc-conf
sudo sysv-rc-conf
```

As you can see in Figure 18.5, the sysv-rc-conf presents a table showing you everything that's running in the background on your Pi. You can use the keyboard arrow keys to move around the table and press the Spacebar to toggle services on and off for the various run levels (that's what the 1–6 and S mean in the column headings). Press Q to quit the tool and to return to your shell prompt.

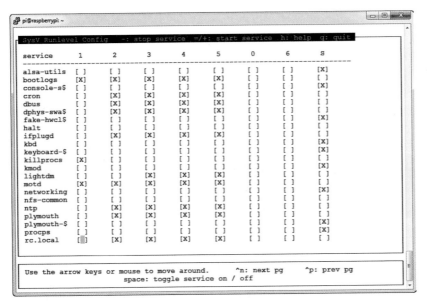

FIGURE 18.5 You can use the sysv-rc-conf utility to disable startup services that you don't need and thereby boost the performance and security of your Raspberry Pi.

Naturally, you'll want to do some research on what these services actually do before you disable anything.

NOTE: ABOUT RUN LEVELS

The Linux operating system can run under different levels of processing priority and system access; these are reasonably called *runlevels*. Only one runlevel is executed on each startup. For instance, runlevel 0 is used to shut down the system. Runlevel 1 is single-user mode, in which only one login is allowed. Runlevels 2–5 are for multi-user access. Runlevel 6 denotes a reboot condition. You can read more about runlevels on Wikipedia at http://is.gd/rTxU06.

By default, the Raspberry Pi reserves six connection terminals (called tty connections) for simultaneous user access. If you are the only person logging into your Pi, then you can save some system resources by reducing this number.

Open the /etc/inittab file in nano or your favorite text editor:

```
sudo nano /etc/inittab
```

Now, comment out the unneeded terminal reservations. On my system, I want to reserve only two connections:

```
1:2345:respawn:/sbin/getty --noclear 38400 tty1
2:23:respawn:/sbin/getty 38400 tty2
# 3:23:respawn:/sbin/getty 38400 tty3
# 4:23:respawn:/sbin/getty 38400 tty4
# 5:23:respawn:/sbin/getty 38400 tty5
# 6:23:respawn:/sbin/getty 38400 tty6
```

Save your changes, close the /etc/inittab file, and reboot your Pi to put the change into effect.

Finally, you can make use of a couple cool apt-get parameters to keep your installed software and repository cache nice and tidy:

```
sudo apt-get autoremove
sudo apt-get autoclean
```

The autoremove parameter removes software packages that were installed by apt-get automatically to satisfy dependencies for some installed and potentially removed software.

The autoclean parameter clears out your local repository of retrieved package files, removing only package files that are no longer valid.

For best performance, I suggest you run those apt-get statements on your Pi every month or so.

## CHAPTER 19

Raspberry Pi and Arduino

As novel as the Raspberry Pi is, you have to remember that the device is simply another Linux box. In other words, the Raspberry Pi, despite its tiny form factor, contains all of the trappings of a full-sized computer.

As nice as it is to have the input/output (I/O) capability of a full Linux machine, you still must deal with the unfortunate side-effect of system overhead. At times the Raspberry Pi cannot get out of its own way, so to speak, when you need it to perform certain tasks, especially those tasks that require precise timing and calibration.

Single-board microcontrollers like the Arduino are perfect for more direct applications that do not require intensive computing or graphical processing power. For instance, what if you wanted to design a wearable microcontroller that lights up a row of LEDs sewn into a leather jacket?

As it happens, the Arduino team in Italy made just such a microcontroller: the Lily Pad (http://is.gd/80MqhJ).

What if you wanted to design and control a robot composed almost entirely of paper? Again, the Arduino community has you covered with the PAPERduino (http://is.gd/5dSyJd).

As far as I am personally concerned, the Arduino is totally awesome. The good news is that you can integrate the Arduino with the Raspberry Pi in a number of different ways.

In this chapter I start by providing a bit of history on the Arduino platform. Next, I dig into the Arduino Uno, the reference Arduino model. I then show you how to get the Uno and Raspberry Pi talking to each other and exchanging data. Finally, I introduce an excellent Arduino clone that is about as easy to use with the Pi as any hardware I've ever seen.

Shall we get started?

Introducing the Arduino

Massimo Banzi, the cofounder of Arduino, said it best when he stated, "The Arduino philosophy is based on making designs rather than talking about them."

The Arduino is a family of single-board microcontrollers that are completely open source. Yes, you heard me correctly: In contrast to the Raspberry Pi, which contains Broadcom-proprietary pieces and parts, the Arduino boards are completely open to the public.

This open source approach and Arduino's Creative Commons-based licensing means that anybody in the world can design (and sell) their own Arduino clones. The only licensing aspect that the Arduino team feels strongly about is that clones must not contain the entire word "Arduino" in their names; this term is reserved for the official boards.

Speaking of boards, the Arduino team, which is based in Italy, manufactures and sells a large number of them. Check them out at the Arduino website at http://is.gd/6P6WSe; here are some of my favorite models:

- **Arduino Uno (http://is.gd/SiSPg1):** This is their most popular board and is ideal both for learning as well as for practical application.
- **Arduino Mega 2560 (http://is.gd/VPzRMr):** This is a much bigger board intended for more comprehensive projects.
- **Arduino LilyPad (http://is.gd/80MqhJ):** This is a cute, wearable microcontroller (see Figure 19.1).
- **Arduino Esplora (http://is.gd/YeZIIg):** This board, which is also pictured in Figure 19.1, has lots of I/O possibilities and is focused squarely at game system designers.

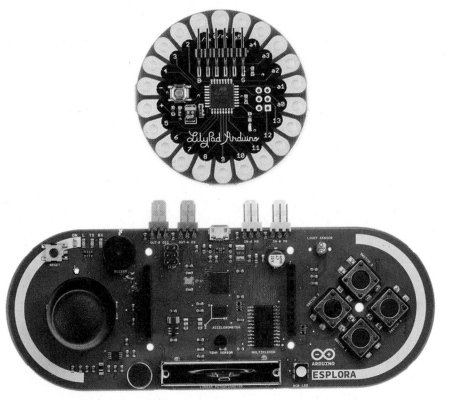

FIGURE 19.1 Arduino LilyPad top, and Arduino Esplora bottom

Okay. Thus far we understand that Arduino represents a family of microcontroller boards and that their hardware is open source and anybody can download the schematics and build Arduino clones. Stepping back for a moment, what exactly is a "single-board microcontroller"?

A single-board microcontroller is a microcontroller that is soldered onto a single printed circuit board (PCB). Going further, a microcontroller, such as the 8-bit ATmel AVR used in the Uno, is an integrated circuit DIP chip that contains a processor core, a small amount of memory, and the capability to communicate with I/O peripherals that are located elsewhere on the PCB.

Note the decided absence of a video controller; microcontrollers have no built-in graphics capability, nor do they contain an operating system. This means you must program and control an Arduino from outside the Arduino, such as from a connected personal computer.

As a matter of fact, the mechanics of connecting an Arduino to the Raspberry Pi are exactly the same as those that govern connecting an Arduino to your Windows, OS X, or Linux-based PC. But I am getting a bit ahead of myself.

For me, the two coolest things about Arduino are (a) its analog inputs; and (b) shields.

The Arduino's analog input pins mean that you can take analog measurements—for instance, temperature, volume, and so on...and convert them to digital values for processing. This means you can use the Arduino to interact with the real world, which is largely analog.

NOTE: ANALOG VERSUS DIGITAL SIGNALS

The reason quantities such as volume are analog is because their values change constantly over time over a wide range of values. By contrast, in the digital world there are only two values: 0 and 1, off and on, low and high, and so forth. When you convert an analog signal to digital, you try to approximate the wave-like pattern of analog to the up and down zig-zag of digital. The more data you feed into the translation, the more faithful the digital signal is of its analog counterpart (and vice versa).

A shield is an add-on board that extends the functionality of the Arduino. Typically, shields connect directly on top of the Arduino's I/O pins...like a soldier holding a shield in front of him, actually.

Shields put the Arduino board on some serious steroids, let me tell you. It seems that either the Arduino team or a third party has developed an add-on shield for every conceivable computing (or noncomputing) purpose. For instance, take a look at this representative smattering of popular Arduino shields:

- **Ethernet Shield (http://is.gd/PU2D0c)**: Gives your Arduino a wired Ethernet connection (see Figure 19.2).
- **Wi-Fi Shield (http://is.gd/URoZZ9)**: Gives your Arduino connectivity to 802.1 b/g wireless networks.
- **GSM Shield (http://is.gd/JFB4NV)**: Gives your Arduino access to carrier networks (you also need a cellular service carrier's SIM card).
- **Relay Shield (http://is.gd/P63Pb4)**: Gives your Arduino the ability to control devices that use higher voltage circuits (perfect for home automation projects).
- **Proto Shield (http://is.gd/nkmvNI)**: Gives you the ability to create your own shield from scratch—this is simply an unpopulated PCB with header connectivity to Arduino.

FIGURE 19.2 Arduino Uno with Ethernet shield attached on top

The AlaMode that you learn how to use toward the end of this chapter is actually an Arduino shield.

Before we proceed into studying the Arduino Uno in great detail, let me return to the concept of the powerful Arduino community. You don't have to have lots of money and manufacturing resources at hand to build your own Arduino.

As it happens, you can design and create an Arduino either from individual parts or by making use of several starter kits. Here, have a look:

- **DIY Arduino (http://is.gd/70D1hc)**: This project is pretty novel, but is likely to require quite a bit of time to undertake.

- **Build an Arduino on a Breadboard (http://is.gd/AeVfiP)**: Believe it or not, it is feasible to build yourself an Arduino clone for all of $5 in parts.

- **Adafruit Arduino Starter Pack (http://is.gd/7qFOJw)**: This kit includes an Uno R3 and myriad doo-dads for your experimenting and project prototyping pleasure.

- **Arduino Starter Kit (http://is.gd/NS4ff2)**: This project kit, which includes an Uno R3, is sold by the Arduino team directly.

Okay, then. Let's drill into the Arduino Uno so we can begin to appreciate what the Arduino can do in Technicolor.

Digging into the Arduino Uno

The word Uno, as you probably already know, means one in Italian, and the Arduino Uno is called Uno to denote its association with the upcoming Arduino 1.0 PCB. According to the Arduino website, they are positioning the Uno and the Arduino 1.0 as the reference PCBs going forward.

As of this writing, the Arduino team has made three revisions to the Uno, with each subsequent revision adding features. Thus, I advise you to purchase an R3 board if you have anything to say about it.

You can easily tell which Uno revision you have by turning the board over and looking for the Rx label (where x is the revision number). For example, the Revision 3 board is labeled R3.

Now what's on the front of the Uno? Figure 19.3 provides an annotated picture for you, but let me break the components down in a bit more detail:

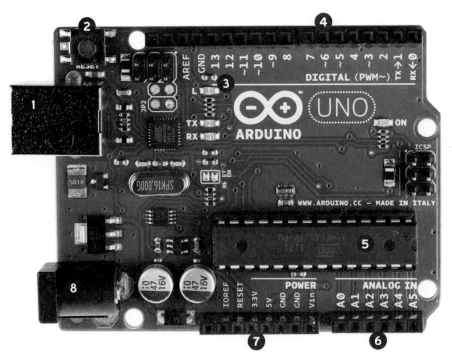

FIGURE 19.3 Arduino Uno PCB, front view

1: USB for data/power

2: Reset button

3: Pin #13 "blink" LED

4: Digital input/output pins

5: 7V-12V DC power

6: 3.3V and 5V power pins

7: Analog input pins

8: ATmega microcontroller

- **ATmel ATmega328P AVR Microcontroller**: The "brains" of the Arduino. The chip has a 16 MHz clock speed and 32KB of flash memory.
- **14 digital output pins**: Six of these pins support Pulse Width Modulation, or PWM. PWM enables you to take digital input and produce analog output. For instance, you can use PWM to dim LEDs instead of turning them on and off completely.
- **6 analog inputs**: This is how you get analog sensor data (think of volume, temperature, brightness, motion, and so on) into the Arduino.

- **USB-to-serial chip**: The ATmega16U2 enables the Uno's USB bus to send and receive serial data.

- **Power supply**: The Arduino itself operates at 5V, which means you need to be careful when you work with the 3.3V Raspberry Pi. The good news here is that the power output pins on the Uno support both voltages. Incidentally, the 3V3 pin supplies up to 50mA of power.

- **Reset**: You can use this tactile button switch to reboot the Arduino.

Power-wise, the Arduino Uno can accept power either through its dedicated power supply or through USB. If both are connected, the Uno defaults to using the dedicated 5V power supply.

NOTE: WHAT'S A "WALL WART"?

The Arduino Uno uses a standard 9V–12V, 250mA or more, AC-to-DC power supply with a 2.1mm plug. You can learn more about Uno-compatible power supplies by visiting Arduino Playground at http://is.gd/9WYgSK. The reason why these power supplies, which are ubiquitous in today's portable electronics age, are called wall warts is because of the plug itself. As you doubtless know and much to your chagrin, the bulky transformer tends to block additional ports in your wall power receptacle or surge protector. Hence the disparaging term.

As I stated earlier, you can program the Arduino from an external computer. Sure, you could connect your Uno to your Windows or Mac computer, but in this book I focus squarely on the Raspberry Pi. Therefore, let's now turn your attention to how you can link these two wondrous devices.

Connecting the Arduino and the Raspberry Pi

One of the fundamental lessons about physical computing that you should have picked up thus far is that any given task has several different valid methods of approach.

Some ways of solving a problem may be more efficient than others; yet others are more or less expensive to undertake. As long as you're satisfied with the end result, there is no single, best way.

The reason I mention this is that there exist several methods for connecting the Arduino Uno to your Raspberry Pi. Again, you have some ways that are more or less efficient (and dangerous!) than others. Let me describe for you how some of these different types of connections work.

Connecting the Raspberry Pi GPIO Pins to the Arduino Serial Pins

This method requires the use of a voltage divider or logic level converter (buy one from Sparkfun at http://is.gd/Ws16r8) to manage the 3.3V/5V voltage difference between the two devices.

The advantage to this approach is that you free up the Raspberry Pi's USB port for another use. The disadvantage is, as I just said, you must account for the voltage difference; doing this typically involves the introduction of a breadboard to host the logic level converter and jumper wires.

If you're brave, you can study Oscar Liang's tutorial on connecting the Raspberry Pi and Arduino Uno via serial GPIO, found at http://is.gd/I2QY7T.

Connecting the Raspberry Pi GPIO Pins to the Arduino I2C

This connection method does not require a logic level converter, as long as you configure the Pi as a master device and the Arduino as a slave device. Here are step-by-step instructions, again from Oscar Liang: http://is.gd/XBDg13.

The advantage to this method is that you free up both the USB bus as well as the serial pins on both devices. The disadvantage is that configuration is difficult. For instance, if you mess up the master/slave I2C communication between the devices you can easily fry your Pi with an overvoltage.

Connecting the Raspberry Pi to the Arduino via USB

Ah yes...simplicity itself. This is the cleanest connection method insofar as you can literally plug the Arduino Uno into one of the Pi's USB ports and access the Uno as a serial USB device.

The issue with this connection method, naturally, is one of power. If you have a wall wart power supply for your Arduino, you're all set. Another solution is to plug your Uno into a powered USB hub that is in turn connected to the Pi's USB port. That's actually the method I use.

Connecting the Raspberry Pi to the Arduino via a Shield or Bridge Board

You now know what Arduino shields are, and it should come as no surprise to you that developers have taken it upon themselves to create Arduino-Raspberry Pi connection shields.

One of the most popular shields in this category is the Ponte (http://is.gd/nAvtEi). This shield is in a "currently experimental" state as of this writing, but it looks like a promising project.

Imagine the possibilities of stacking a Ponte on top of a Raspberry Pi, an Arduino on top of the Ponte, and another Arduino shield stacked on top of the Arduino!

Connecting the Raspberry Pi to an Arduino Clone

The AlaMode shield and the Gertboard are both Arduino clones. Although the AlaMode is an Arduino clone that fits on top of the Raspberry Pi's GPIO header like any traditional shield, the Gertboard's ATmega microcontroller is just one of a number of widgets soldered into this multipurpose experimentation board. I show you how to use the AlaMode in this chapter, and you learn more about the Gertboard in Chapter 20, "Raspberry Pi and the Gertboard."

Simply connecting the Arduino and the Raspberry Pi is only half the battle. You also have to take a look at the software side of the equation. This involves three discrete tasks:

- Configuring the Pi to recognize the Arduino
- Installing the Arduino IDE software
- Developing sketches on the Pi and uploading them to the Arduino

NOTE: ALL ABOUT SKETCH

In Arduino nomenclature, a sketch is nothing more than a script file that constitutes your program source code. Sketch scripts, which will come up again, in the section "Task: Install and Configure Arduino IDE on the Raspberry Pi," later in this chapter, are plain text files with an .ino extension that are readable in any text editor.

I believe the term sketch is meant to denote the programmer's ability to quickly and easily sketch out his or her ideas in code and to be able to test the code immediately on a connected Arduino device.

Understanding the Arduino Development Workflow

You already know that the Arduino has no operating system of its own. I myself look at the Arduino as basically a dumb terminal. You can compose a set of instructions on a remote device, and then upload that script to the Arduino, where the script is stored in flash memory.

Immediately, the Arduino begins executing what's in the script. Assuming the script is error-free and the Arduino does not suffer a hardware problem, the device will dutifully perform that work, theoretically forever.

Even if you press the hardware button to reset the Arduino, the currently loaded sketch continues to play. Any microcontroller worth its salt is all about reliably performing a single purpose.

The Arduino team created a piece of (surprise!) open source software called Arduino IDE (http://is.gd/UXSYgL) that you can use to program the Arduino. The software is free and is available for Windows, OS X, and Linux.

Specifically, Arduino IDE is a Java application and is based on the Processing programming language. Processing (http://is.gd/c6fUpT) is a C-type, object-oriented programming language that was created for those in the visual design community to teach the fundamentals of software development.

The tricky piece with installing the Arduino IDE on the Raspberry Pi is the Java Runtime Environment (JRE) requirement. Remember what I said earlier in the book about Java's heaviness and the Pi's tendency to choke on Java code? Yeah, that.

We're going to rely on a splendid installation recipe that was developed by Kevin Osborn of the Bald Wisdom blog (http://is.gd/7RYaFm). Oh, Kevin is also on the AlaMode development team (http://is.gd/UvXxMF).

TASK: INSTALL AND CONFIGURE ARDUINO IDE ON THE RASPBERRY PI

Perform the following actions on your Pi from a shell prompt. This procedure also assumes that you are using the Raspbian "Wheezy" Linux distribution and not Adafruit's Occidentalis or another custom distro.

In this chapter, I use USB and a powered hub to connect the Arduino Uno and my Raspberry Pi Model B board. A schematic diagram of my setup is shown in Figure 19.4.

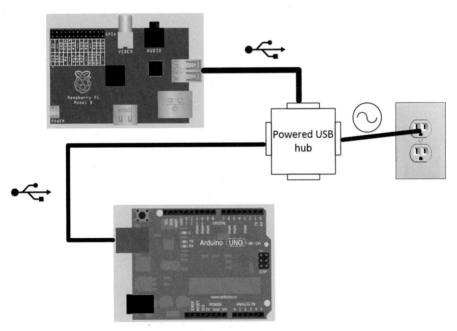

FIGURE 19.4 My Arduino-Raspberry Pi physical topology

1. First, make sure that your Pi's firmware and system software are up-to-date. We originally covered this subject in Chapter 4, "Installing and Configuring an Operating System."

2. Install the Arduino IDE; this step is simplicity itself:

```
sudo apt-get install -y arduino
```

3. Had Kevin not created a shell script for us, you would have had a dozen or more tedious configuration steps to undertake to force the Raspberry Pi to recognize the Arduino as a serial device (recall that the Uno includes an ATmel USB-to-serial chip; the Raspberry Pi has no such onboard convenience on its end).

   If you want to understand all of the steps involved anyway, simply open Kevin's script in a text editor and study away!

```
wget https://github.com/wyolum/AlaMode/blob/master/bundles/AlaMode-setup.tar.
gz?raw=true -O AlaMode-setup.tar.gz
tar -xvzf AlaMode-setup.tar.gz
cd AlaMode-setup
```

4. Note the reference to AlaMode. Kevin wrote this script with a dual-purpose in mind; namely, to configure the Pi to communicate both with the official Arduino Uno as well as the third-party AlaMode shield. I get to the AlaMode later; for now, run the setup script:

```
sudo ./setup
```

5. Once setup completes, you can reboot the Pi, or you can simply run the following command to initialize the system:

```
sudo udevadm trigger
```

6. To start the Arduino IDE, you need to be in a graphical environment, so make sure you fire up LXDE.

7. You will find an Arduino IDE shortcut in the Programming folder in the LXDE programs launcher, or you can run the following simple command from LXTerminal:

```
arduino
```

   After a few moments, you should see the Arduino IDE interface as shown in Figure 19.5.

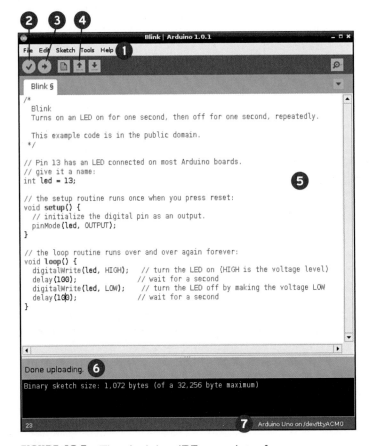

FIGURE 19.5   The Arduino IDE user interface

As you can see in this figure, I annotated the Arduino IDE user interface. Let me take a moment to teach you the major controls:

1: Menu system. Here you can find all the Arduino IDE commands nested in categories.

2: Verify. Here you can validate the syntax of your sketch.

3: Upload. Here you can transfer the current sketch to the Arduino's flash memory.

4: New, Open, and Save, respectively. These buttons represent typical file management commands.

5: Sketch area. This is where you compose your Arduino source code.

6: Status area. This is where you receive feedback from the Arduino IDE with regard to code validation, compilation, and transfer.

7: Device information. This verifies that the Uno is linked, and reveals the serial port to which the Uno is connected.

NOTE: WHAT PROGRAMMING LANGUAGE DOES THE ARDUINO IDE USE?

It's actually rather confusing to determine exactly what programming language you're using when you write sketches using the Arduino IDE. According to the Arduino website, the Arduino IDE employs its own open source programming language called, reasonably enough, the Arduino programming language.

Syntactically and functionally, the Arduino programming language is a simplified version of C/C++ and actually uses the avr-gcc C compiler. While we are on the subject, you might want to check out Simon Monk's on-the-money book *Programming Arduino: Getting Started with Sketches* (http://is.gd/tdnYTw) if you're looking for guided, end-to-end instruction on using the Arduino programming language.

Before you load up and run your first sample sketch, let's take a moment to ensure that Arduino IDE is properly configured. First, point your mouse to the menu bar and click Tools, Board and make sure that IDE is set to Arduino Uno.

Second, click Tools, Serial Port and ensure that IDE is set to /dev/ttyACM0. Incidentally, ACM stands for abstract control model and refers to the ability to transmit old school serial data over the newer-school USB bus.

TASK: RUNNING THE "BLINK" SAMPLE SKETCH

The Arduino IDE includes a library of starter sketches. In the same way that "Hello world" is the first program many of us write when we learn a new language, the Blink sample sketch is by far the most popular sketch to try initially with the Arduino.

Before undertaking this task, ensure that your Raspberry Pi and Arduino are both powered on and connected.

1. In Arduino IDE, click File, Examples, 1. Basics, Blink. This loads the sketch into the IDE (see Figure 19.6).

2. Click the Upload button or click File, Upload Using Programmer to send the sketch to your Uno. The status bar reads Compiling sketch, Uploading, and then Done uploading when the transfer is complete.

Look at the surface-mounted LED marked L on the Uno board. Do you see it blinking once per second?

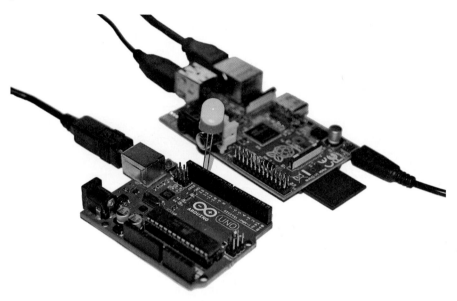

FIGURE 19.6 Testing out the Blink test sketch. Note that I've added an extra LED (you'll learn how to do that, too).

As shown here, you can put some more light on the subject by plugging a standalone LED into digital output pins 13 and ground. Before you do this, however, make sure you insert the longer leg (positive, or anode) into pin 13, and insert the LED's shorter leg (negative, or cathode) into the GND pin.

Electrical current always flows from the anode to the cathode. The anode leg is made longer than the shorter simply as an easy way for us to determine which LED leg is which.

Because the surface-mount LED is also wired to pin 13, you should see both LEDs flash in unison.

You can learn a lot, exert control over the Arduino, as well as have some fun by analyzing the Blink sketch code line-by-line. Here's the full code from the Blink test file; we'll then analyze each major statement:

```
/*
  Blink
  Turns on an LED on for one second, then off for one second, repeatedly.

  This example code is in the public domain.
*/
```

```
// Pin 13 has an LED connected on most Arduino boards.
// give it a name:
int led = 13;
// the setup routine runs once when you press reset:
void setup() {
  // initialize the digital pin as an output.
  pinMode(led, OUTPUT);
}
// the loop routine runs over and over again forever:
void loop() {
  digitalWrite(led, HIGH);    // turn the LED on (HIGH is the voltage level)
  delay(1000);                // wait for a second
  digitalWrite(led, LOW);     // turn the LED off by making the voltage LOW
  delay(1000);                // wait for a second
}.
```

First of all, notice that the sketch includes two functions, setup() and loop(). The setup() procedure initializes a particular digital pin as a signal output.

How do we know we want pin 13? It's right here:

```
int led = 13;
```

This statement creates a variable named led that can accept only integer (whole number) values. Furthermore, this line initializes the value of the led variable to 13. The 13 denotes pin #13 on the Arduino. You can find pin 13 by examining the digital pins on top of the Arduino Uno PCB. (Hint: pin 13 is located immediately to the right of the GND interface.)

```
pinMode(led, OUTPUT);
```

The pinMode()function accepts two arguments; first you pass in the variable, which is pin 13, and OUTPUT marks the pin for outgoing signal/current.

The loop() procedure does exactly what you think it does: it performs whatever actions you specify in the code block an infinite number of times. As previously stated, an Arduino runs its currently loaded sketch forever, unless and until you send it a new sketch to run. Singleness of purpose, remember?

```
digitalWrite(led, HIGH);
```

The digitalWrite()function puts the target LED in an on state (HIGH) or an off state (LOW). This is binary data we're talking about.

```
delay(1000);
```

The delay function accepts an integer value in milliseconds. A value of 1000 means that the LED will stay on for 1 second.

Thus, the blink sketch does nothing more than cycles LED 13 on and off every second... forever. As an experiment, change the delay value from 1000 to 100 and reupload the sketch to your Uno. You should see the LED blink much faster.

NOTE: HOW TO CLEAR OUT YOUR ARDUINO

If you want to stop the Blink sketch, or if you want to clear any of the Arduino's current actions, start a new, blank sketch file that contains the following lines:

```
void setup(){};
void loop(){};
```

Verify the sketch and then upload it to your Uno. I actually saved the sketch on my Pi and named it Clear for easy reuse.

Fading an LED

In this example with the Arduino Uno you test the use of analog output and PWM to gradually fade an LED.

Following is an inventory of what you need to have on hand to complete this experiment:

- Arduino Uno
- Breadboard
- 220 ohm resistor
- LED

You can study Figure 19.7 to learn the wiring schematic for this test.

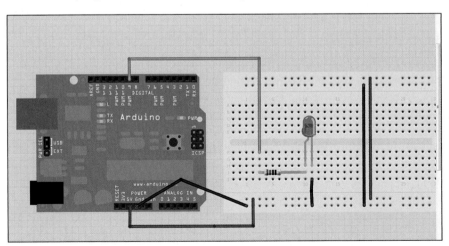

FIGURE 19.7 Wiring diagram for our Arduino PWM LED test

With respect to that wiring diagram, please take into account the following notes:

- The (+) power rail is for incoming power.

- The (-) power rail is for outgoing power (ground).

- The Uno board in my schematic is from an earlier PCB release. As long as you choose a digital output pin labeled "PWM" or "~" you are good to go.

TASK: RUNNING THE FADE SKETCH

You now know the basics of uploading and running Arduino sketches, so this second example procedure should feel more natural for you to undertake.

1. When your wiring is complete, fire up Arduino IDE and click File, Examples, 1. Basics, Fade.

2. Go ahead and upload the sketch. While you do so, keep an eye on the TX and RX surface-mounted LEDs on the Uno PCB. You'll notice them flash a bit as the Ardunio sends and receives data, respectively.

3. The end result of this test is you see your LED move gradually from full intensity to off and then back again, giving the illusion of a fading effect.

4. To learn what's going on under the hood, you can again examine the Fade sketch source code.

```
int led = 9;
int brightness = 0;
int fadeAmount = 5;
```

This code defines three variables. The led variable denotes PWM pin number 9. The initial bulb brightness (PWM value) is set to 0, which equals off. The fadeAmount variable controls how granular or choppy the fade effect is, as you see in a moment.

The setup() procedure simply specifies pin #9 as output, as you saw in the previous exercise. The loop() procedure is much more interesting here, though.

```
analogWrite(led, brightness);
```

The analogWrite()function sets the brightness of pin 9 to its current value, which we already know is 0.

```
brightness = brightness + fadeAmount;
```

Here you increment the brightness value by the fadeAmount value. You can tweak the script by adjusting either of these variables.

```
if (brightness == 0 || brightness == 255)
    {fadeAmount = -fadeAmount ; }
```

The previous expression, when stated in human terms, says, "If the LED brightness reaches either the maximum brightness of 255 or minimum brightness of 0, then reverse the direction of fadeAmount."

In other words, while the LED increases its brightness from 1 to 254, the fadeAmount value is positive.

When the brightness value hits 255, however, the fadeAmount maintains its value, but reverses its direction, thus becoming negative. This is what allows the LED to "fade" from full brightness down to off. Get it?

Using the AlaMode

Now that you have gotten to know an official Arduino board to a good degree of depth, let's close out this chapter by looking briefly at one of the most popular Arduino clones, at least from the perspective of a Raspberry Pi enthusiast.

The AlaMode (http://is.gd/mm0Kfd) is an Arduino clone/shield that fits directly on top of the Raspberry Pi's GPIO header pins. I show you an annotated close-up of the AlaMode in Figure 19.8.

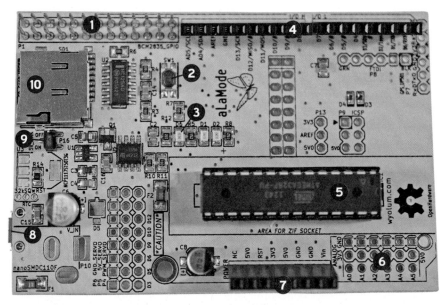

FIGURE 19.8   The AlaMode perfectly fits the Pi's GPIO header pins.

1: GPIO

2: Reset button

3: "Blink" LED

4: Digital I/O pins

5: Micro SD card slot

6: Power source jumper

7: Micro USB power input

8: ATmega microcontroller

9: Power pins

10: Analog input pins (unpopulated)

The AlaMode includes an impressive array of features besides its ease of connectivity to the Raspberry Pi:

- **DS3234 Real Time Clock**: Neither the Raspberry Pi nor the Arduino Uno has a battery-backed RTC. This is an excellent feature because you may want to run tasks that require precise timing and the Pi does not include a real-time clock "out of the box."

- **Micro-SD Card Slot**: You can perform data logging without having to access a network. Again, this is a tremendous convenience for certain projects.

- **Power Flexibility**: You can power the AlaMode directly from the Raspberry Pi through the GPIO, or you can plug in a wall wart or connect a traditionally DC battery.

The AlaMode's microprocessor is the same ATmega328P that you have on the Uno. AlaMode connects to the Pi as an I2C slave device and performs 5V-3.3V buffering. The AlaMode also includes a general-purpose blink LED on pin 13, just like the Arduino.

You can purchase the AlaMode through several channels; here are a couple:

- Maker Shed: http://is.gd/6eMMnC
- Seeed Studio: http://is.gd/fFFDnQ

Be sure to snag a copy of the user manual, too: http://is.gd/2bniUh.

TASK: GETTING THE ALAMODE UP AND RUNNING

You would have been surprised, had you not read this book, that the AlaMode ships without I/O headers installed. This means you need to solder them into the AlaMode yourself to gain the PCB's full functionality.

Unfortunately, soldering deserves a chapter unto itself. As a consolation, here are some top-notch resources that I picked for you that will teach you everything you need to know (actually, soldering the headers to the AlaMode is a 15-minute procedure; it truly is no big deal):

■ Make Video Podcast, Soldering Tutorial: http://is.gd/MALror

■ Curious Inventor, How to Solder Correctly, and Why (video): http://is.gd/XIcaVx

■ Electronix Express, Better Soldering: http://is.gd/9FYfXL

■ Soldering is Easy Comic Book: http://is.gd/aNcuNQWith regard to the software configuration, as long as you've performed the steps given in the earlier procedure "Task: Install and Configure Arduino IDE on the Raspberry Pi," you've completed most of the work.

Let's sweep up the shavings together now:

1. Power off your Raspberry Pi.

2. If you want to power the RTC, insert a CR1632 battery into the associated clip on the AlaMode.

3. Gently push the AlaMode onto the Raspberry Pi's GPIO pins. Make sure to line up the Pi GPIO with the AlaMode in the correct orientation; you can see the Pi-AlaMode sandwich (as well as some unsoldered headers) in Figure 19.9.

FIGURE 19.9 The Raspberry Pi, the AlaMode, and an unsoldered I/O header

4. Now about power. You can configure the AlaMode to receive its power directly from the Pi's GPIO header (assuming the Pi receives at least 1A of current on its own) or from a wall wart power supply.

In this tutorial, let's power the AlaMode directly from the Pi. To do this, you must first set the 5V_Link jumper to ON as shown in Figure 19.10.

FIGURE 19.10 The AlaMode's 5V_Link jumper is set by default to receive power through the Raspberry Pi GPIO.

The good news is that the AlaMode ships with the jumper set this way by default, so this is a verification step rather than a configuration step. By contrast, if you want to use an external power supply, you must move the jumper so that it covers the other pin, setting the switch to OFF.

Other than that, you can load up Arduino IDE on the Raspberry Pi and send sketches to the AlaMode in the very same way you did with the Arduino Uno.

5. In Arduino IDE, click Tools, Board and select AlaMode. Finally, click Tools, Serial Port and ensure that port /dev/ttyS0 is selected. Happy experimenting!

## CHAPTER 20

Raspberry Pi and the Gertboard

Fans of the Gertboard claim that the device is the ultimate expansion board for the Raspberry Pi. Having spent quite a bit of time with the Gertboard, I can tell you that there is indeed substance to that claim.

Gert van Loo (pronounced *van LOW*) is a Broadcom engineer who also happens to be one of the principal designers of the Raspberry Pi. As I'm sure you can correctly guess, Gert also invented the Gertboard.

Formally defined, the Gertboard is an expansion board, also called a daughterboard, that connects to the Raspberry Pi GPIO headers and gives you instant access to a tremendous variety of input/output options.

You can look at the Gertboard as an activity center or toybox with which you can experiment with motors, switches, buttons, and even an onboard Arduino microcontroller.

Basically, the Gertboard extends the Raspberry Pi to the real world, allowing you to sense temperatures, detect sounds, drive motors, and so forth.

Take a look at Figure 20.1 and the following descriptions for a tour of the PCB's major components.

FIGURE 20.1   The Gertboard is a multipurpose expansion board for the Raspberry Pi.

1: 12 LEDs

2: 3 momentary button switches

3: Motor controller

4: 6 open collector driver inputs

5: GPIO

6: Atmel ATmega chip

7: 10-bit Analog-to-Digital and 8-bit Digital-to-Analog converters

The L6203 motor controller drives brushed DC physical motors, including servos and steppers.

The ULN2803a open collector drivers enable you to turn devices on and off, especially those that use a different voltage than the Gertboard itself or those that use more current than the Gertboard can supply.

The Atmel ATmega 328P AVR microcontroller gives you built-in Arduino prototyping capability. One important note about this on-board Arduino chip is that it runs at 3.3V instead of the standard 5V Arduino voltage.

The MCP4801 Analog-to-Digital (A2D) and MCP3002 Digital-to-Analog (D2A) converters, as you'd expect, enable you to process both analog as well as digital audio signals. This hardware is especially useful if you want to, for instance, detect an input volume for an alarm system.

You learn more about the LEDs, button switches, and GPIO pins momentarily. In the meantime, you are probably wondering where you can purchase a Gertboard.

Originally, the Gertboard was sold in an unassembled state that required soldering. Fortunately, Farnell/Element 14 now sells an assembled model that is ready to rumble for $49 USD, as of this writing. Go purchase a Gertboard at http://is.gd/mnQiHJ.

The assembled Gertboard is called "revision 2" and is physically much different (and improved) from revision 1. For instance, the revision 1 didn't have an Arduino-compatible Atmel controller, and the board was much more cluttered than the revision 2 model.

Gert himself posted an excellent walkthrough of the Gertboard revision 2 changes in a video on his YouTube channel at http://is.gd/ArQfMK.

Anatomy of the GPIO

The Raspberry Pi General Purpose Input/Output (GPIO) is a 26-pin expansion header that is marked on the PCB as P1 and employs a 2x13 copper pin strip. I provide the GPIO pinout in Figure 20.2.

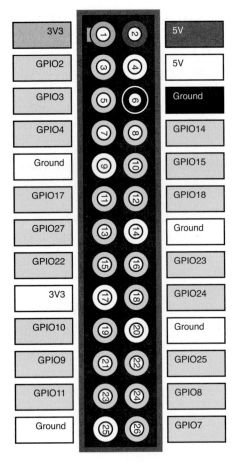

FIGURE 20.2   The Raspberry Pi GPIO pinout

The pins that you see here are arranged into four basic functionality groups:

- 2 +3.3V voltage pins
- 2 +5V voltage pins
- 5 Ground pins
- 17 GPIO pins (access to I²C, SPI, and UART)

Raspberry Pi PCBs fabricated after September 2012 are called "revision 2" boards. As it happens, the Foundation changed the function of three GPIO pins between revision 1 and revision 2; for more information, read the article at the Embedded Linux Wiki (http://is.gd/CNI2JC).

The bottom line, friends, is that unless you're using Ethernet, the GPIO represents the only way to interface your Raspberry Pi with other device hardware. The GPIO header pins are amazingly flexible; they can be reprogrammed to support input or output, and they can be selectively enabled or disabled.

One important note that I've mentioned before that bears repeating: although the GPIO has two 5V pins, the GPIO voltage levels are tolerant only of 3.3V signals, and there is not built-in overvoltage protection on the Raspberry Pi. Consequently, if you are not careful you can fry your Pi.

The 5V pins at P1-02 and P1-04 on the Model B board support a maximum current draw of 300mA.

The official name for the Raspberry Pi GPIO is "The GPIO Connector (P1)." Individual pins on the GPIO header are referred to with the P1 prefix; for instance, P1-01, and so forth.

Okay—enough background information. Let's connect the Gertboard to your Raspberry Pi and begin some serious experimentation!

Connecting Your Gertboard and Raspberry Pi

The Gertboard connects to the Raspberry Pi pin-for-pin by using the GPIO headers. You can either (carefully) mount the Gertboard directly on top of the Pi board, or you can use a 26-pin ribbon cable. You can see a Gertboard/Raspberry Pi sandwich in Figure 20.3.

FIGURE 20.3   The Gertboard connects to the Raspberry Pi board by using the GPIO header.

It's important to be careful when you connect the Gertboard and the Raspberry Pi. You need to line up every male GPIO pin on the Pi with the corresponding female socket on the underside of the Gertboard. To make this process simpler, you can use the plastic standoffs that ship with the Gertboard to create a more stable surface for the board.

If you want to use a GPIO ribbon cable to make the connection, then you need not only the cable (https://www.modmypi.com/gpio-accessories/ribbon-cables-and-connectors/ raspberry-pi-GPIO-assembled-rainbow-ribbon-cable-and-connectors), but also a female-to-male converter (https://www.modmypi.com/gpio-accessories/gpio-header-extenders/ 26-Pin-GPIO-Shrouded-Box-Header).

NOTE: ON JUMPERS AND GENDER

With jumper wires, or any connection cable for that matter, the male end of one wire or cable inserts into the female (recessed) end of another wire or cable. As you might surmise, this gender-related nomenclature is intimately associated with human reproduction.

Now about power. As long as your Raspberry Pi power supply can give at least 1A, you can power the Gertboard directly from the Pi. The incoming power from the Pi operates at 5V, and the Gertboard uses either 5V or 3.3V depending on which components you use.

In practice, we make connections within the Gertboard and between the Gertboard and the Raspbery Pi by using the straps and jumpers included in the assembled Gertboard kit.

The female-to-female straps enable you to connect pins located in different locations on a board or between boards. By contrast, shunt-type jumpers connect immediately adjacent pins. I show you what these connectors look like in Figure 20.4.

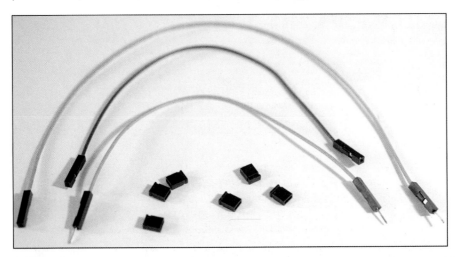

FIGURE 20.4   Traditional shunt-type jumpers along with several types of jumper wire straps (male-to-female, female-to-female, and male-to-male)

Now I'm about to save you a lot of troubleshooting time: Make sure to place a jumper on the two J7 header pins, as shown in Figure 20.5.

FIGURE 20.5 You can add 3.3V of power to the Gertboard components by adding a jumper to the two pins of header J7. The J7 header has three pins, and the jumper covers the top two when the Gertboard is viewed "right-side up."

By adding this jumper to the Gertboard, you allow 3.3V of power to flow to all of the Gertboard's components.

Installing the Testing Software

Gert wrote a suite of small C programs that test various Gertboard components. However, we've focused on Python in this book, and I'll continue that here by using the Python Gertboard code modules.

NOTE: FOR YOU C PROGRAMMERS

To be honest, Gert's C test suite accesses the Gertboard hardware more directly than does the Python test suite. If you want to download the C code, feel free to do so at http://is.gd/PuS9FU.

The Python test suite was created by Alex Eames of Raspi.TV. Alex wrote the software in Python 2.7, but you'll recall that Raspbian includes both Python 2 as well as Python 3.

To access Alex's code in its entirety, you need to have the RPi.GPIO and WiringPi libraries installed. Raspbian, as of September 2012, includes the RPi.GPIO library by default, but you need to install WiringPi (http://wiringpi.com/) yourself. Take a moment and run the following commands from a shell prompt on your Raspberry Pi:

```
sudo apt-get update && sudo apt-get upgrade
sudo apt-get install python-dev python-pip
sudo pip install wiringpi
```

NOTE: GPIO LIBRARIES

The RPi.GPIO and WiringPi libraries both perform the same actions; namely, allowing programmatic access to the Raspberry Pi GPIO header. Alex Eames wrote versions of his Python Gertboard test suite to accommodate both libraries because neither library offers a fully complete set of capabilities. You can read a nice discussion of RPi.GPIO versus WiringPi at the Raspberry Pi forums at http://is.gd/SGAFNp and http://is.gd/wJ1qfb.

TASK: ENABLING SPI ON YOUR RASPBERRY PI

Some of the scripts, notably atod.py, dtoa.py, and dad.py, require that you enable Serial Peripheral Interface (SPI) on your Pi. You can take care of that prerequisite by performing the following steps from a shell prompt on your Raspberry Pi (the Gertboard does not have to be connected at this point although there is no harm done if it is:

1. Open the raspi-blacklist.conf file.

```
sudo nano /etc/modprobe.d/raspi-blacklist.conf
```

2. Comment out the following line in the configuration file with a # such that it appears like this:

```
#blacklist spi-bcm2708
```

Performing this action prevents SPI from being disabled on your Raspberry Pi.

3. Save your work, close the file, and reboot:

```
sudo reboot
```

4. After the reboot, install Git and the Python SPI wrapper:

```
sudo apt-get install git
git clone git://github.com/doceme/py-spidev
cd py-spidev
sudo python setup.py install
```

Although the previous procedure was a bit tedious, the joy that you'll receive in accessing the Gertboard's D-to-A and A-to-D converters should make your effort worthwhile.

Now let's turn our attention to loading the Python Gertboard test suite.

TASK: INSTALLING THE PYTHON GERTBOARD TEST SUITE

Perform the following steps from a shell prompt on your Raspberry Pi:

1. Get the library from Raspi.tv, unpack the ZIP file, and navigate into the extracted folder:

```
cd
wget http://raspi.tv/download/GB_Python.zip
unzip GB_Python.zip
cd GB_Python
```

2. Run a directory listing to view the listing of Python scripts and then check out the informative README text file:

```
ls -l *.py
nano README.txt
```

3. Now that the scripts are unpacked, you can run any of them by invoking sudo privileges and the Python 2 interpreter. For instance, try out the RPi.GPIO-based LED testing script:

```
sudo python leds-rg.py
```

The script given here won't do anything yet because you have to wire up the Gertboard appropriately. We get to that in just a moment.

You can see that Alex wrote two versions of each script file. The -rg file name designation denotes the RPi.GPIO library, while the -wp suffix represents the WiringPi library. Here is a brief rundown of each script's purpose; note that I removed the -rg and -wp suffixes and added a generic xx placeholder to the script names for clarity:

- **atod.py**: Tests the analog-to-digital converter
- **butled-xx.py**: Tests the button switches and LEDs

- **buttons-xx.py**: Tests the button switches
- **dad.py**: Tests both analog-to-digital and digital-to-analog
- **dtoa.py**: Tests the digital-to-analog converter
- **leds-xx.py**: Tests the LED switches
- **motor-xx.py**: Tests the motor
- **ocol-xx.py**: Tests the relay switches
- **potmot.py**: Tests the analog-to-digital switch and the motor

Testing the LEDs

Your first test lights up the light emitting diode (LED) panel on the Gertboard. You can use either the leds-rg.py or the leds-wp.py script in this case; the RPi-GPIO and WiringPi libraries can both handle this exercise with no problem.

If you examine the source of the leds-rg.py script, you see that Alex documented it pretty well, including giving abbreviated directions for how to wire up the Gertboard. Check out the first few lines:

```
pi@raspberrypi ~/GB_Python $ sudo python leds-rg.py
These are the connections for the Gertboard LEDs test:
jumpers in every out location (U3-out-B1, U3-out-B2, etc)
GP25 in J2 --- B1 in J3
GP24 in J2 --- B2 in J3
GP23 in J2 --- B3 in J3
GP22 in J2 --- B4 in J3
GP21 in J2 --- B5 in J3
GP18 in J2 --- B6 in J3
GP17 in J2 --- B7 in J3
GP11 in J2 --- B8 in J3
GP10 in J2 --- B9 in J3
GP9 in J2 --- B10 in J3
GP8 in J2 --- B11 in J3
GP7 in J2 --- B12 in J3
(If you don't have enough straps and jumpers you can install
just a few of them, then run again later with the next batch.)
When ready hit enter.
```

In this documentation, the Bs and GPs refer to marked locations on the Gertboard. Don't worry—Gert himself provides excellent board diagrams that show exactly where you should put the jumpers and straps for each of the tests in his suite.

To view the wiring diagram for this test (and for all the other tests, for that matter), download and view the Gertboard user manual at http://is.gd/dOWlUd. To save you the download, you can see the wiring schematic in Figure 20.6 as well.

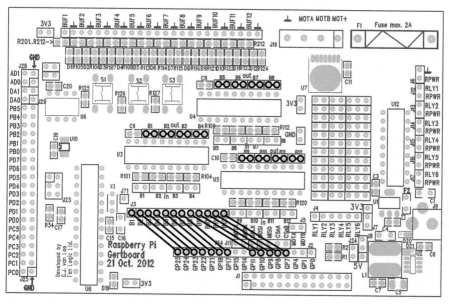

FIGURE 20.6   The wiring diagram for Gertboard LED test

Figure 20.7 shows you what my lab setup looks like while the test is running.

FIGURE 20.7 My Gertboard/Raspberry Pi setup for the LED test. I hooked up only the first six LEDs.

You need to use sudo when you run the script, and remember that these scripts were written in Python 2, so make sure you invoke the Python interpreter instead of python3.

When you run the script, observe that the Gertboard's LEDs flash in three discrete patterns. Don't worry if you don't have enough straps and jumpers to wire all 12 of the LEDs; hook up as many or as few as you want, and the script will ignore any unconnected LEDs.

NOTE: POWER ON OR POWER OFF?

I'm sure you're wondering, "Do I need to unplug the Raspberry Pi every time I adjust a strap or jumper on the Gertboard?" Although the politically correct answer is yes, I have experienced no strange behavior, nor have I damaged either the Gertboard or the Raspberry Pi by moving from test to test while keeping both boards powered up and online.

To customize the behavior of the LED flashes, open the script in nano or your favorite text editor and play around with the led_drive() function; this is the primary "engine" of the script.

Specifically, the reps parameter defines how many times to run the test. The multiple parameter specifies whether or not to switch off an LED before proceeding to the next one. Finally, the direction parameter defines, well, the directionality of the LED actions, either left-to-right or right-to-left.

Thus, to run the test 10 times in the reverse direction, change the led_drive definition to match the following:

```
led_drive(10, 0, ports_rev)
```

Testing Input/Output

The LED test that we just finished tested only the Gertboard's ability to render output. By contrast, the button/LED test enables you to send the output of a button press to a particular LED as input.

For this test, use the butled-rg.py script because, as of this writing, the test works only with the RPi.GPIO libraries. The file name here is illustrative; it stands for but(ton) and, of course, LED.

The wiring for this test is more straightforward than for the LED test (by which I mean the test requires fewer straps and shunt jumpers). You can view the schematic in Figure 20.8.

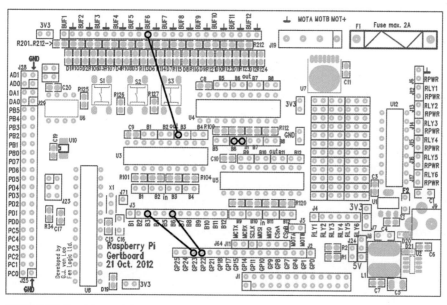

FIGURE 20.8 The wiring diagram for Gertboard Input/Output test

Run the test by issuing the command

```
sudo python butled-rg.py
```

You see output on your screen as you press the S3 button on the Gertboard, and you also see the BUF6 LED respond to each button press. The screen output cycles between binary 11 (button not pressed, 3.3V voltage (HIGH state) and binary 00 (button pressed, 0V current, LOW state).

I challenge you to figure out how to activate the other two button switches and link them to two additional LEDs. Have fun!

A Couple Quick Breadboarding Exercises

In Chapter 3, "A Tour of Raspberry Pi Peripheral Devices," you learned what a breadboard is and how important the tool is in prototyping hardware. I wanted to devote a bit of space in this book to showing you a couple of breadboarding experiments that you can undertake with your Raspberry Pi.

Both experiments execute as simple a task as possible: lighting an LED. However, I want to show you how you can run the experiment by using a breadboard and the naked Raspberry PI GPIO pins as well as how you can do the same thing with the Adafruit Pi Cobbler that I told you about in Chapter 3.

Accessing the GPIO Pins Directly

Let's begin by looking at a shopping list of parts that you need to complete this introductory prototyping experiment:

- 1 standard breadboard
- 1 LED
- 1 resistor in the 270–330 ohm range
- 2 male-to-female jumper straps

You can find all of these parts at your local RadioShack or electronics shop. Alternatively, I've gotten *a lot* of mileage out of electronics prototyping kits. To that point, here are some suggestions I've found useful:

- Sparkfun RedBoard Breadboard Kit (http://is.gd/FbYgZ5)
- RadioShack Breadboard and Jumper Wire Kit (http://is.gd/XGEBWz)
- MakerShed Mintronics Survival Pack (http://is.gd/UMcR9O)

NOTE: SOMETIMES "DISH" IS NOT RELATED TO GOSSIP

Adafruit sells the wonderful Pi Dish ($22.50, http://is.gd/n4TmjO) that enables you to secure a Raspberry Pi and a standard breadboard in an attractive, clear, and sturdy package. An accessory like this makes Raspberry Pi breadboarding infinitely cleaner and easier than dealing with free-floating PCBs, straps, and jumpers.

TASK: LIGHTING AN LED DIRECTLY FROM THE GPIO

In this procedure, pull current from GPIO pin #1 at 3.3V and feed it through an LED mounted on the breadboard. The resister is mounted in the same circuit; the higher the resistance value, the dimmer the light shines. If your resister is too small, you will burn out the LED bulb. Finally, the resistor bleeds off excess current to the ground rail on the outside rim of the breadboard.

1. Turn off your Raspberry Pi.

2. Carefully press the resistor and male-to-female jumper straps in place. You can see the wiring diagram in Figure 20.9.

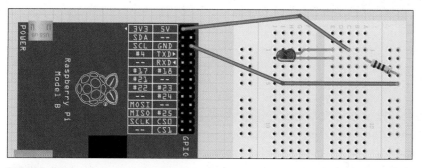

FIGURE 20.9   The wiring diagram for our first breadboarding experiment

NOTE: HOW I CREATED MY WIRING DIAGRAMS

To create these nifty wiring diagrams, I used a wonderful piece of open source software called Fritzing (http://is.gd/i17V7t). With a few mouse clicks, you can document your prototypes in an eye-appealing and accurate way. What's more, Adafruit publishes a Github repository with Raspberry Pi and Pi Cobbler Fritzing parts (http://is.gd/ZLCdPk). Great stuff!

3. The LED has a longer leg and a shorter leg. The longer leg is the positive terminal and goes in the same breadboard row as the 3.3V jumper. The shorter leg is the negative terminal and goes in the same row as the resistor.

4. Power on the Raspberry Pi. If everything is hooked up correctly, you should see the LED light up immediately.

Let's extend this experiment such that you gain control over the illumination state of the LED. To do this, you again access the WiringPi libraries.

Physically, all you have to do is relocate the female jumper pin currently plugged into GPIO #1 to GPIO pin #11.

With that done, you need to perform a little bit of housekeeping with the WiringPi libraries to send commands to the Raspberry Pi GPIO pins directly.

Download and then compile the WiringPi executable code:

```
cd
git clone git://git.drogon.net/wiringPi
cd wiringPi
git pull origin
./build
```

Awesome! Now you can stay in the current directory and issue GPIO commands to your heart's content. For instance, try the following:

```
gpio mode 0 out
gpio write 0 1
gpio write 0 0
```

You should find that the gpio write 0 1 command turned the LED on and that the gpio write 0 0 line turned the LED off. It's like you're turning on a faucet: when you send 3.3V out of programmable GPIO #11 pin into the LED, the electrical energy is consumed and emitted as light.

Accessing the GPIO Pins via the Pi Cobbler

The Adafruit Pi Cobbler (http://is.gd/B1U0bq) represents a more elegant way to make the Raspberry Pi GPIO pins accessible to you and your projects. Instead of wiring individual jumpers from the GPIO pins to the breadboard, you can break out from the GPIO header directly to the breadboard and then access GPIO from there.

Adafruit sells the Pi Cobbler either unassembled or assembled; I leave it up to you and your tolerance for pain (just kidding) in deciding which product to purchase. In addition to the Pi Cobbler breakout IC, you also get a ribbon cable.

To mount the Pi Cobbler, you first must connect the cable to the Cobbler board itself. Adafruit was nice enough to add a notch in the 26-pin ribbon cable, so it is impossible to insert the cable into the Cobbler incorrectly.

You can potentially get into trouble by inserting the other end of the ribbon cable into the Raspberry Pi GPIO header, though. Locate the colored edge wire of the ribbon cable; this is pin #1 and needs to be inserted into the GPIO on the side closest to the SD card slot and where P1 is marked on the Raspberry Pi board.

You also need to take care to insert the Pi Cobbler in the breadboard such that the Cobbler straddles the bridge or center line. Be sure to press the Cobbler all the way into the breadboard. You can see a close-up of my installed Pi Cobbler in Figure 20.10.

FIGURE 20.10 The Pi Cobbler is simultaneously connected to the breadboard and the Raspberry Pi and serves to extend the GPIO to the board. The markings above each Cobbler pin match each corresponding GPIO pin on the Raspberry Pi.

Whew! Now that you have that out of the way, what can you actually *do* with the Pi Cobbler?

Well, look closely at the perimeter of the Cobbler—you should see markings that correspond to each of the 26 pins of the Raspberry Pi GPIO.

You can easily repeat your initial LED experiment by inserting male-to-male (note that you need male ends when your connection begins and ends on the breadboard) jumper straps in the same breadboard row as the corresponding GPIO header.

The only change I make, as noted in Figure 20.11, is that I used a GPIO ground pin on the same side of the breadboard as the power pin.

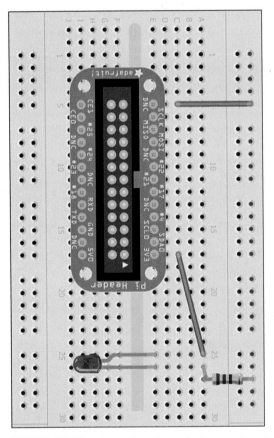

FIGURE 20.11 A replication of the earlier breadboarding experiment, this time by using the Pi Cobbler breakout board

Programming the Atmel Microcontroller

That long, 28-pin DIP you see soldered in next to the left of the Gertboard logo on the PCB is none other than an Atmel AVR ATmega328p microcontroller, the same chip that you experimented with in the previous chapter.

NOTE: SUBJECT TO PARTS AVAILABILITY...

Take a moment to read the identification information on the surface of your Gertboard's Atmel chip. Gert states that due to parts availability, the Gertboard may be equipped with either the ATmega 328 or 168.

This means you can perform Arduino experiments by compiling sketches on the Raspberry Pi and sending them to the ATmega directly. Now then, recall that the Arduino operates at 5V, and the Raspberry Pi operates at 3.3V. The upshot of this situation for our purposes is that the Gertboard's ATmega runs at a slower clock speed (12MHz instead of 16MHz).

Thus, if you plan to reuse some of your sketches from Chapter 19, "Raspberry Pi and Arduino," you need to adjust any references to +5V or you may very well fry the Gertboard's ATmega chip. You also may need to adjust the timing of your sketch code to account for the Gertboard's slower clock speed.

TASK: PREPARING YOUR ARDUINO ENVIRONMENT

I covered installing the Arduino IDE in Chapter 19. However, I want to give you the complete procedure now in case you haven't done any work with the Arduino UNO yet.

Perform the following tasks from a terminal prompt on your Raspberry Pi:

1. Start by downloading and installing the Arduino IDE:

```
sudo apt-get install -y arduino
```

2. Use AVRDUDE to help you upload your Arduino sketches to the AVR microcontroller on the Gertboard.

```
cd /tmp
wget http://project-downloads.drogon.net/gertboard/avrdude_5.10-4_armhf.deb
sudo dpkg -i avrdude_5.10-4_armhf.deb
sudo chmod 4755 /usr/bin/avrdude
```

3. The bad news is that there are several steps involved in completing the Raspberry Pi-Gertboard Arduino setup. The good news is that Gordon Henderson, a British computer consultant with jaw-dropping expertise with the Raspberry Pi, Gertboard, and Arduino environments, graciously created a script that automates these steps. Check out Gordon online at http://is.gd/7SPmYJ.

```
cd /tmp
wget http://project-downloads.drogon.net/gertboard/setup.sh
chmod +x setup.sh
sudo ./setup.sh
```

4. After the script completes, you are prompted to reboot your Raspberry Pi. Do that.

5. Initialize the ATmega chip. Make sure your Gertboard is installed and attach the jumper wires as shown in Figure 20.12.

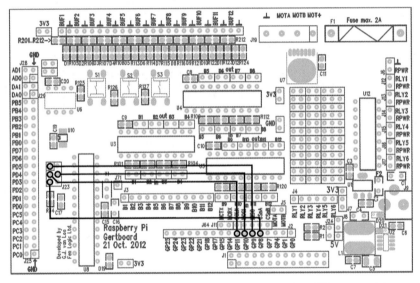

FIGURE 20.12 The wiring diagram to initialize the ATmega chip on the Gertboard

6. Issue the "magic" initialization command:

```
avrsetup
```

You see the following output:

```
Initialising a new ATmega microcontroller for use with the Gertboard.

Make sure there is a new ATmega chip plugged in, and press
.. 1 for an ATmega328p or 2 for an ATmega168: 1
```

7. If you have the ATmega 328 on your Gertboard, type **1** and press Enter. (Type **2** if you have the ATmega168.) If all goes well, you'll see the following output:

```
Initialising an ATmega328p ...
Looks all OK - Happy ATmega programming!
```

8. You're almost home-free. You just have a bit of configuration to do in the Arduino IDE itself. First, fire up the IDE:

```
arduino
```

9. In the Arduino IDE, Click Tools, Board and select Gertboard with ATmega328(GPIO) from the flyout menu.

10. Next, click Tools, Programmer and select the Raspberry Pi GPIO option.

To test functionality by using the built-in Blink sketch, you first need to attach a jumper strap from location PB5 on the left side of the Gertboard to one of the buffered LED outputs. (I use BUF6 as a matter of practice.)

After you've connected the wire, go back to the Arduino IDE and click File, Examples, 01.Basics, Blink, and then click the Upload button (or click File, Upload using Programmer) to send the sketch to the ATmega chip.

You should see the BUF6 LED begin to blink slowly.

Final Thoughts, and Thank You

Well, congratulations! You've reached the end of this book. We've certainly come a long way, haven't we?

By way of a take-home message, I encourage you to stay current with all things Raspberry Pi-related by plugging into and participating in the community. The most direct entry points into the Raspberry Pi community are as follows:

- **Official Raspberry Pi Community Forum**: http://is.gd/6nBR5Z
- **Google+ Raspberry Pi Forum**: http://is.gd/jGajWj
- **Stack Exchange: Raspberry Pi**: http://is.gd/hWh8EK
- **RPi Community Links at eLinux.org**: http://is.gd/sN9O4g

If you completed even the majority of tasks in this volume, then you have an excellent grasp of the fundamentals of hardware and software hacking. Should you have any questions, or if you want to share your experiences all the way, I'm all ears. You can reach me directly at tim@timwarnertech.com.

Thank you so very much for purchasing this book and for learning more about the Raspberry Pi. It is people like you who keep information technology relevant and growing over time.

Happy hacking!

INDEX

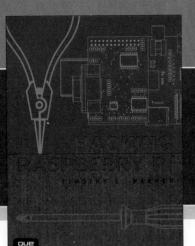

FREE
Online Edition

Your purchase of *Hacking Raspberry Pi*® includes access to a free online edition for 45 days through the **Safari Books Online** subscription service. Nearly every Que book is available online through **Safari Books Online**, along with thousands of books and videos from publishers such as Addison-Wesley Professional, Cisco Press, Exam Cram, IBM Press, O'Reilly Media, Prentice Hall, Sams, and VMware Press.

Safari Books Online is a digital library providing searchable, on-demand access to thousands of technology, digital media, and professional development books and videos from leading publishers. With one monthly or yearly subscription price, you get unlimited access to learning tools and information on topics including mobile app and software development, tips and tricks on using your favorite gadgets, networking, project management, graphic design, and much more.

Activate your FREE Online Edition at
informit.com/safarifree

STEP 1: Enter the coupon code: IDYAOVH.

STEP 2: New Safari users, complete the brief registration form.
Safari subscribers, just log in.

If you have difficulty registering on Safari or accessing the online edition,
please e-mail customer-service@safaribooksonline.com